GeoGuide

Series Editors

Wolfgang Eder, GeoCentre-Geobiology, University of Göttingen, Göttingen, Niedersachsen, Germany

Peter T. Bobrowsky, Geological Survey of Canada, Natural Resources Canada, Sidney, BC, Canada

Jesús Martínez-Frías, CSIC-Universidad Complutense de Madrid, Instituto de Geociencias, Madrid, Spain

Axel Vollbrecht, Geowissenschaftlichen Zentrum der Universität Göttingen, Göttingen, Germany

The GeoGuide series publishes travel guide type short monographs focussed on areas and regions of geo-morphological and geological importance including Geoparks, National Parks, World Heritage areas and Geosites. Volumes in this series are produced with the focus on public outreach and provide an introduction to the geological and environmental context of the region followed by in depth and colourful descriptions of each Geosite and its significance. Each volume is supplemented with ecological, cultural and logistical tips and information to allow these beautiful and fascinating regions of the world to be fully enjoyed.

Hans-Rudolf Wenk · Elizabeth Wenk

Discovering the Unique Geology of the Bergell Alps

Springer

Hans-Rudolf Wenk
Department of Earth and Planetary
Science
University of California, Berkeley
Berkeley, CA, USA

Elizabeth Wenk
School of Biological, Earth
and Environmental Sciences
The University of New South Wales
Sydney, NSW, Australia

ISSN 2364-6497   ISSN 2364-6500 (electronic)
GeoGuide
ISBN 978-3-031-30737-9   ISBN 978-3-031-30738-6 (eBook)
https://doi.org/10.1007/978-3-031-30738-6

© The Editor(s) (if applicable) and The Author(s), under exclusive license to Springer Nature Switzerland AG 2023

This work is subject to copyright. All rights are solely and exclusively licensed by the Publisher, whether the whole or part of the material is concerned, specifically the rights of translation, reprinting, reuse of illustrations, recitation, broadcasting, reproduction on microfilms or in any other physical way, and transmission or information storage and retrieval, electronic adaptation, computer software, or by similar or dissimilar methodology now known or hereafter developed.
The use of general descriptive names, registered names, trademarks, service marks, etc. in this publication does not imply, even in the absence of a specific statement, that such names are exempt from the relevant protective laws and regulations and therefore free for general use.
The publisher, the authors, and the editors are safe to assume that the advice and information in this book are believed to be true and accurate at the date of publication. Neither the publisher nor the authors or the editors give a warranty, expressed or implied, with respect to the material contained herein or for any errors or omissions that may have been made. The publisher remains neutral with regard to jurisdictional claims in published maps and institutional affiliations.

This Springer imprint is published by the registered company Springer Nature Switzerland AG
The registered company address is: Gewerbestrasse 11, 6330 Cham, Switzerland

Preface

The Bergell Alps offer not only a magnificent landscape with impressive mountains that attract many tourists, but also they document a unique geological history, first explored by Bernhard Studer (1851) and since then studied and documented in many publications. In this book, we try to connect visitors with scientific curiosity with remarkable geological features observed in the Bergell Alps, such as a young granite that intrudes into an older stack of nappes (thrust sheets) that represent the European continental plate at the base overthrust by the African plate, causing gradients in temperature and pressure that can be explored by looking at outcrops of rocks in the field.

While there are several books introducing the geology of northwestern America to a broad public with observations in the field, there are few in Europe, though with equally fascinating geological features. We admired this approach in the book of Glazner and Stock (2010): Geology Underfoot in Yosemite National Park. Their book is not just meant for geologists, but also for readers with curiosity in physical and environmental sciences who love to connect with the outdoor world and would like to learn how our landscape evolved.

We dedicate the book to Konrad Springer (1925–1997) who in his career as publisher not only tried to advance science but also to connect laymen and experts through books and journals. He had a background in biology and became attracted to Bergell geology on several excursions with Rudy Wenk (Fig. 1). The Springer *GeoGuide* series looks like an excellent place to advance this connection. We started with a first brief guide many years ago (Wenk 1985), and this book is a much more coherent and comprehensive expansion.

We acknowledge Remo Maurizio from Vicosoprano (1933–2017) who spent his life exploring and describing the natural history—animals, plants, rocks and minerals—of the Bergell Alps (e.g. Maurizio and Weibel 1982). We took many

Fig. 1 Konrad Springer in the Bergell, summer 1975

unforgettable fieldtrips together, including one with Konrad Springer whom he fondly remembered because they were contemporaries taking courses in the biological sciences at the University of Zurich.

We are also appreciative for the hospitality by many people in the Bergell who made us to feel in this region like a second home, families like Bischoff, Capadrutt, Fagetti, Ganzoni, Giacometti, Giovanoli, Hunkeler, Pasini, Picenoni, Ruinelli and many others.

We are most grateful to Walter Hunkeler from Soglio who stimulated this project and let us use some of his pictures. His booklet "Warum der Piz Badile so jung ist" (Hunkeler 2022) is an introduction, awakening the reader's curiosity about Bergell geology, and we are following up on it. We were also inspired by the suggestion of Mazzoleni et al. (2013) to create a geopark in the region Masino–Bergell because of its unique features.

We acknowledge contributions from colleagues such as Hans Bänninger, Francesco Bedogne (1942–2012), Andreas Fasciati (EWZ), Andrea Galli (ETH), Marco Giacometti (Centro Giacometti, Stampa), Michael Graf (Lenzerheide), Susan Ivy-Ochs (ETH), Guido Mazzoleni, Attilio Montrasio (CNR Milano), Eli Müller (Tourismus Bregaglia), Adrian Pfiffner (University of Bern), Bruna Ruinelli and Jakob Messerli (Museum Ciäsa Granda in Stampa), Enrico Sciesa, Yiming Zhang (Berkeley) and Comune di Bregaglia, who gave permission to include extraordinary images and access to their work.

We are also appreciative to SwissTopo for supporting part of the fieldwork and providing access to their complex digital map system, as well as published documents. Thank you Andrea Bühler.

On a more personal note, Rudy Wenk would have never become engaged in earth sciences without his father Eduard Wenk, who took him already as a young boy on many geological excursions and made him enthusiastic about minerals and rocks. He also accompanied our family on many hikes in the Bergell mountains. His wife Julia, also with a background in geology, helped edit the book, and our daughter Elizabeth, second author, knows the challenging Bergell peaks better than many mountaineers.

Lastly, we gratefully acknowledge Springer Verlag and particularly Editor Doris Bleier and her production team for making this project possible. We hope this book will inspire many to explore geological history on their hikes, and perhaps some may follow up with books linking science curiosity with aspects of geology in the field.

Berkeley, USA
Sydney, Australia

Hans-Rudolf Wenk
Elizabeth Wenk

References

Glazner AF, Stock GM (2010) Geology Underfoot in Yosemite National Park. Mountain Press, Missoula, 300 pp

Hunkeler W (2022) Warum der Piz Badile so jung ist. Soglio Produkte, Castasegna, 113 pp

Maurizio R, Weibel M (1982) Die Mineralien des Bergells. Mineralienfreund 20(4):81–100

Mazzoleni G, Garotuno M, Paganoni A, Mazzoni G (2023) Dalla Via Geoalpina al progetto di Geoparco del Granito nella "Regione Geologica" del Masino-Bregaglia. G & T Viaggio nella geologica d'Italia, Regione Lombardia. Conf Congr Naz Geol e Turismo 5

Studer B (1851) Geologie der Schweiz (Band 1). Stämpfli, Bern/Schulthess, Zürich, pp 485

Wenk H-R (1985) Introduction to the geology of the Bergell Alps with guide for excursions. Jber Natf Ges Graubünden 103:29–90

Contents

1	**Introduction**		1
	References		5
2	**The Geological Framework**		7
	References		21
3	**Six Highlights of Bergell Geology**		23
	3.1	Minerals and Rocks in the Bergell	23
	3.2	Bergell Granite	37
	3.3	Microstructure of Rocks, Their Composition and Formation Conditions	40
	3.4	The Recent History: Glaciers and Moraines	44
	3.5	Landslides	48
	3.6	Mining, Quarrying and Water	52
		3.6.1 Lavez (Olivine-Talc Schist)	52
		3.6.2 Lime Production	53
		3.6.3 Quarries of Gneiss and Gravel Pits	56
		3.6.4 Metal Ores	56
		3.6.5 Water/Electricity	58
	References		62
4	**Geological Excursions in the Bergell Alps**		65
	4.1	Easy Hikes	69
		4.1.1 Alp Cavloc: Tectonic Nappes and Contact Rocks (Map 1, waypoints #a1–#a10) (2–5 h)	69

	4.1.2	Maloja: Torre Belvedere, Moraines, Peat Bogs and Glacial Mills (Map 2, waypoints #b1–#b3) (1–2 h)	72
	4.1.3	Val Maroz: Greenschists, Bündnerschiefer, Travertine, Serpentinite (Map 4, waypoints #c1–#c8) (4–6 h)	73
	4.1.4	Albigna Hut: Bergell Granite, with Cable Car (Map 5, waypoints #d1–#d4) (4–8 h)	76
	4.1.5	Maira-Vicosoprano: Diversity of Bergell Rocks in the River (Map 5, waypoints #e1–#e2) (1–2 h)	78
	4.1.6	Piuro-Borgonuovo, Palazzo Vertemate-Franchi, Explore Lavez and Glacial Features in a Walk from Prosto (Map 3, waypoints #f1–#f8) (2–4 h)	80
	4.1.7	Chiavenna, Outcrops of Tambo Granite in the Liro River, Cimaganda Landslide (Map 3, waypoints #g1–#g7) (2–3 h)	83
4.2	Longer Excursions (1–2 days)		86
	4.2.1	Bergell Granite and Contact Zone, Val Forno Glacier, Forno Hut, Monte del Forno, Pillow Structures, Muretto Pass (Map 6, waypoints #h1–#h10) (2 days)	86
	4.2.2	Maloja, Pass Lunghin, Piz Lunghin: Nappes, Serpentinite, Triassic Marble (Map 7, waypoints #i1–#i5) (5–8 h)	91
	4.2.3	Plaun da Lej-Grevasalvas-Plaun Grand, Fuorcla Grevasalvas, Plaun dal Sel, Blaunca: Higher Nappes, Radiolarite, Old Granites (Map 7 waypoints #j1–#j10) (6–8 h)	93
	4.2.4	Casaccia-Val Maroz-Val da la Duana-Piz Duan-Cadrin-Löbbia-Soglio: Pennine Nappes (Avers, Suretta, Tambo) (Map 4, waypoints #k1–#k10) (2 days)	96
	4.2.5	Roticcio-Val Furcela-Val da Cam-Piz Cam: Young Diabase Dikes, Stratigraphy of Suretta Nappe, Mn-Silicates (Map 3, waypoints #l1–#l10) (6–8 h)	100

	4.2.6	Val Albigna, Pass da Casnil, Piz Casnil or Piz Bacun, Passo Cacciabella Sud, Pizzo Eravedar. Bergell Granite, Pegmatite and Aplite Dikes, Xenolith Inclusions, Orbicular Granite (Map 5, waypoints #m1–#m10) (2 days)	102
	4.2.7	Lavinair Crusc-Piz Salacina: Calcareous-Silicates in Contact Zone (Map 1, waypoints #n1–#n7) (5–7 h)	105
	4.2.8	Bondo-Ciresc-Lera d'Sura-Denc dal Luf: Al_2SiO_5 Triple Point, Gruf Migmatites, Ultramafics, Moraines (Map 7, waypoints #o1–#o8) (5–8 h)	107
	4.2.9	Bondo-Cugian-Trubinasca-Cap. Sasc Fura (Val Bondasca). Contact Zone of Bergell Granite, Ultramafic Xenoliths, Deformed Granite, View of the 2017 Landslide (Map 7, waypoints #p1–#p5) (7–9 h)	110
	4.2.10	Overview: Via Panoramica from Casaccia to Soglio (Map 3, waypoints #q1–#q6) (4–6 h)	112
	4.2.11	Bondo-Ciresc-Tegiola-Val Codera-Novate Mezzola: Gruf Migmatite, Mylonites, High Temperature Mineral Assemblages, Novate Granite (Map 8, waypoints #r1–#r7) (2 days)	115
	4.2.12	Valmalenco: Serpentinite Near Chiesa, Contact Rocks with Tonalite in Val Sissone (Map 9, waypoints #s1–#s7) (6–8 h)	119
References			121

List of Geological Maps with Itineraries 125

Glossary 145

Index 151

Introduction

The Bergell Alps offer a wild mountain landscape with impressive peaks, steep cliffs and a valley that has been used for both alpine agriculture and a thoroughfare since time immemorial (Fig. 1.1). Romans passed through the valley as a link between Italy and northern Europe, building trails over passes such as the Septimer, Maloja and Julier. The Bergell has also attracted famous people. The physicist Wilhelm Röntgen spent summer holidays at the Hotel Bregaglia in Promontogno where there is still the "Röntgen Zimmer". Around 1894 the Italian painter Giovanni Segantini stayed in the Palazzo Salis in Soglio, painted mountain landscapes and called the Bergell "the threshold to paradise". He inspired the famous local artist family Giacometti from Stampa. Around 1920, the Austrian poet Rainer Maria Rilke lived in Soglio for a few months.

In the villages of the valley many old buildings dating from 1500 to 1900 are preserved. Some are now museums. We have already mentioned the Palazzo Salis in Soglio (1630), but in Bondo there is another Palazzo Salis (1690), today a summer residence for the English branch of Count de Salis. In Castelmur the Palazzo Castelmur, a patrician building from 1723 and long a refuge of the emigrated sugar baker family Castelmur, is now a museum. In Stampa, the Ciäsa Granda from 1581 is the valley museum with exhibitions of art (including Giovanni, Alberto, Diego and Augusto Giacometti), local history, zoology, mineralogy and geology. Vicosoprano, the largest village in the valley, was already inhabited in prehistoric times. In the center is the round Sevelen tower from the thirteenth century. Other treasures include the medieval church San Martino in Bondo, inaugurated in 1250, with wonderful frescoes. We mentioned the Hotel Bregaglia in Promontogno (from 1875) and should not forget the Hotel Palace in Maloja from 1884, also in the municipality of Bregaglia, but more reminiscent of St. Moritz.

There are many opportunities for hikers to spend rainy days in museums such as Ciäsa Granda and Palazzo Castelmur in Stampa, Palazzo Vertemate-Franchi in

Fig. 1.1 View on Val Bregaglia (right) with snow-covered mountains that are mainly granite. Albigna glacier and hydroelectric reservoir in the center. In the background is Lago di Como in Italy. *Source* Virtual Archive of Wild Heerbrugg

Piuro-Prosto, Museo Archeologico della Valchiavenna in Chiavenna and Museo dei Minerali in Sondrio.

Today, the Bergell Valley has fewer farmers and lives largely from tourism, attracting visitors from all over the world to hike and enjoy the wild landscape. Many climb mountains with challenging peaks such as Badile, Sciora, La

1 Introduction

Fiamma, Castello, Forno and Margna. Others may enjoy more relaxing walks and stay in nearby St. Moritz or Pontresina, where luxurious hotels are the essential attraction.

How does this wild landscape in the Bergell come about? It is the geology and in this book we give an introduction how these diverse structures originated in geological history. But it is not only the mountains and rocks that characterize the Bergell Alps. There were also huge landslides such as in 1618 when a landslide destroyed the affluent town of Piuro (Plurs) and killed over 1000 people. In 1673 the village Casaccia, was covered with a stream of debris. Landslides continue: in 2017 a large rock fall from Pizzo Cengalo buried Val Bondasca.

During the Last Ice Age, 115,000–10,000 years ago, the Alps were covered with ice. In the Bergell region only the highest peaks were above the ice sheet. Glaciers eroded the landscape. They produced moraines which now cover a large part of the surface. Today these moraines are mainly forested or used as meadows but impressive relics of the Ice Age remain. Glacial mills in Maloja and Chiavenna are testament to the incredible forces within active glaciers carving the topography beneath them.

Rising above the relatively recent deposits such as moraines, debris fields, rock falls and river alluvium, are mountains with impressive cliffs composed of a variety of rocks. Long before the Last Ice Age vast tectonic activity played a major role in creating the rocks we now see. Through large thrusts, 80 million years ago, caused by the collision of the European and African continents, a stack of nappes (thrust sheets) was formed and associated with it a recrystallization of the rocks under temperature and pressure. For example sandstones and limestones deposited in the ocean transformed to quartzites and marbles. Basaltic lava flows became amphibolites, and remnants of the upper mantle that were once olivine-rich peridotites became serpentinites.

The tectonic movements created the Alps. In the Bergell region temperatures were locally high enough and led to melting and these melts of granitic composition intruded upwards. This happened ~35 million years (My) ago (e.g. Grünenfelder and Stern 1960) and the Bergell Granite is one of the youngest granites in the Alps and even the world.

These rocks formed deep inside the Earth and came to the surface due to erosion by glaciers and rivers over millions of years. The eroded material, pebbles and sand, was deposited in the Po Basin.

The Bergell has some unique geological "triple points" where three parameters meet: (a) One is on the border between the old Europe in the north, the overthrust Africa in the east and a young granite that intruded into this system. (b) Another is

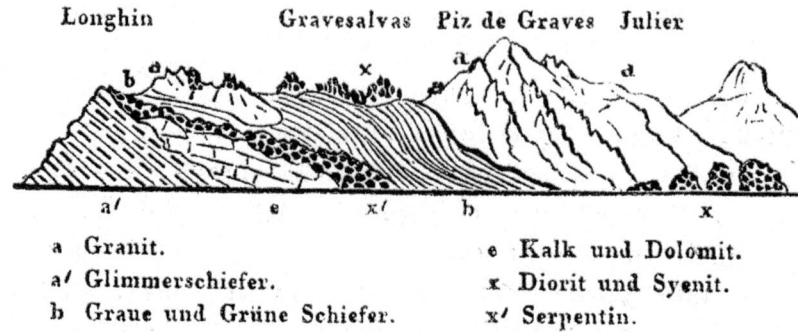

Fig. 1.2 Studer (1851) provides this cross-section of nappes near Maloja

on the Lunghin Pass above Maloja denoting the three directional drainage: northwards into the Rhine and the North Sea; eastwards into the Inn, then the Danube and into the Black Sea; and to the southwest drainage is into the Maira, the Po and the Adriatic Sea. (c) There is a mineralogical triple point above Bondo with an outcrop where three aluminum silicate minerals—andalusite, kyanite and sillimanite (all Al_2SiO_5)—coexist in gneiss and thus define temperature and pressure during crystallization in a temperature–pressure phase diagram.

With this variety of rocks, it is not surprising that the Bergell has fascinated geologists for a long time. In his book on the geology of Switzerland, Studer (1851) devotes extensive discussions on many details, from granite in Val Codera to profiles in Grevasalvas (Fig. 1.2), which show that he himself has studied this hard-to-reach area. Theobald published the first geological map in 1865 and describes the geology of the southern Graubünden with many observations (1866). But it is Staub who was the first to focus on the geology of the Bergell (1918) and in 1921 published the still important detailed map "Geologische Karte der Val Bregaglia (Bergell)". In his book "Bau der Alpen" the Bergell is discussed in detail (1924). This is followed by his geological maps of the Avers (1926) and the Bernina Group (1946). Important contributions also came from Cornelius who explored Val Forno (1913), produced a map of the Err-Julier Group (1932) at the scale 1:25,000 and accompanying explanations (1935, 1950). These early studies attracted Drescher-Kaden who explored Bergell granite and transformations caused by hydrous solutions (e.g. Drescher-Kaden and Storz 1926).

Here we first give an overview of the geology of the Bergell region and how the complicated structures originated. Then we briefly discuss the minerals and rocks, and the conditions under which they crystallized. After that a few words about recent geological history, with glaciers, moraines and landslides.

An important part follows with a series of excursions to familiarize hikers and readers with Bergell geology. Hopefully on your next tour you will not only admire the landscape, cliffs, flora and fauna, but also understand the origins of these unique mountains where the European and African continents meet and a granite intrudes from great depths. It may inspire you to explore further the broad range of geological sciences. And geologists may want to discover more details about this fascinating place with a complex history.

References

Cornelius HP (1913) Geologische Beobachtungen im Gebiete des Fornogletschers (Engadin). Cbl Mineral Geol Paläont 1913:246–252

Cornelius HP (1932) Geologische Karte der Err-Julier-Gruppe, 1:25 000. Geol Spez-Karte Schweiz Nr 97

Cornelius HP (1935) Geologie der Err-Julier-Gruppe: I. Teil. Das Baumaterial. Beitr Geol Karte Schweiz [N.F.] 70(1):321

Cornelius HP (1950) Geologie der Err-Julier-Gruppe. II. Teil: Der Gebirgsbau. Beitr Geol Karte Schweiz [N.F.] 70(2):264

Drescher-Kaden FK, Storz M (1926) Ergebnisse petrographisch-tektonischer Untersuchungen im Bergeller Granit. N Jb Mineral Geol Paläont Beilageband A 54:284–291

Grünenfelder M, Stern TW (1960) Das Zirkon-Alter des Bergeller Granits. Schweiz Mineral Petrogr Mitt 40:253–259

Staub R (1918) Geologische Beobachtungen am Bergellermassiv. Vjschr Natf Ges Zürich 63:1–18

Staub R (1921) Geologische Karte der Val Bregaglia (Bergell), 1:50 000. Geol Spez-Karte Schweiz Nr 90

Staub R (1924) Der Bau der Alpen. Beitr Geol Karte Schweiz [N.F.] 52:1–272

Staub R (1926) Geologische Karte des Avers (Piz Platta—Duan), 1:50 000. Geol Spez-Karte Schweiz Nr 115

Staub R (1946) Geologische Karte der Bernina-Gruppe und ihrer Umgebung im Oberengadin, Bergell, Val Malenco, Puschlav und Livigno, 1:50 000. Geol Spez-Karte Schweiz Nr 118

Studer B (1851) Geologie der Schweiz (Band 1). Stämpfli, Bern/Schulthess, Zürich, 485 pp

Theobald G (1866) Geologische Beschreibung der südöstlichen Gebirge von Graubünden. Beitr Geol Karte Schweiz 3, 1 Serie, inkl. Blatt XX der Geol Karte Schweiz 1:100,000, 1865

The Geological Framework 2

We begin with a few words about the geological history of the Bergell for non-experts. Apologies to geologists: we have simplified some aspects to make it easier to understand for laymen. And a warning for all, geology is a complicated science with many names, chemical formulas and speculations. For some strange names you may want to consult the Glossary at the end of the book. In the Bergell Alps, as is true everywhere in the world, many geological aspects are still not resolved and require further studies. For more complete information about the geology of Switzerland, please consult the literature (e.g. Gnägi and Labhart 2017; Pfiffner 2019).

The Earth has been around for ~4500 million years (My). Geologists have divided this long time period into different episodes with Latin names (Fig. 2.1). Exact ages (scale on the right) are based on isotope analyses and rely on radioactive decay of atoms. For example, in the mineral zircon that crystallizes at high temperatures, there are traces of uranium. A ^{238}U isotope decays over geological times into a lead ^{206}Pb isotope and the uranium/lead isotope ratio can be used to determine the absolute age of the zircon and the rock in which it crystallized. Thirty years ago, zircons were separated and analyses of the isotopes in zircon grains gave an average value. Today, an ion probe can be used to analyze micrometer-sized regions and determine local changes. Zircon is often zoned with an old core and a younger rim. Zircons in the migmatites of Val Codera in the Western Bergell Alps have an old core (300–500 My) and a much younger rim (~30 My) (e.g. Galli et al. 2012). Isotopes such as ^{10}Be and ^{36}Cl can also be used to study the evolution of glaciation in the Alps during the Last Glacial Maximum (e.g. Kamleitner et al. 2022).

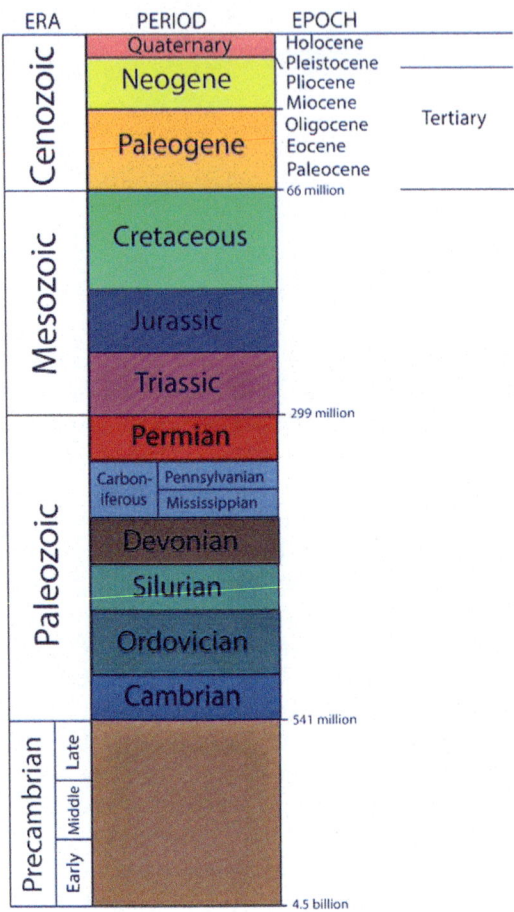

Fig. 2.1 History of the Earth divided into eras, periods and epochs

The oldest igneous rock in the Earth was documented in the Jack Hills of Western Australia (4404 My, Valley et al. 2014), one of the youngest granites is from Mt. Kinabalu in N-Borneo (~7.5 My, Burton-Johnson et al. 2019).

2 The Geological Framework

Bernina granite has an age of ~333 My (e.g. von Quadt et al. 1994), Bergell granite 30–35 My (Grünenfelder and Stern 1960) and Novate granite below the Bergell granite ~20–25 My (e.g. Liati et al. 2000). ^{238}U/^{206}Pb ages describe the crystallization of minerals from a melt. Other isotope ages, e.g. ^{40}K/^{40}Ar in muscovite mica, provide information about cooling history of rocks during the Alpine metamorphism.

Important periods for the Bergell are *Carboniferous* (ancient granitic rocks), *Triassic* (limestone, dolomite and sandstone deposited in the ocean which later recrystallized into marbles and quartzites), *Jurassic* (schists, transformed from clay and shales), *Cretaceous* (thrusting and folding during the continental collision that created the Alpine belt), *Tertiary* (Bergell granite, Alpine metamorphism), *Quaternary* (*Pleistocene*: Ice Age, *Holocene*: Alluvium, landslides).

Geological times have seen major changes in climate. The Carboniferous (~350 My) was dominated by a largely tropical climate with large forests, where wood transformed into coal. In the Triassic (~280 My) there were deposits of conglomerates and sandstones on the continental margin and large coral reefs around islands indicating a relatively dry climate. In the Jurassic period (~200 My) there was sedimentation of clay and limestone in oceanic basins. In the history of the Earth there have been at least five very cold periods where large parts of the world were covered by ice, from the Huronian in the Precambrium (2400 My) to the Pleistocene (the Last Ice Age period 100,000–10,000 years ago).

Already early in Earth's history there have been continents and oceans, like the Precambrian granite in Australia and these continents have been moving over geologic times, as expressed in their magnetic signatures. In 1915 Alfred Wegener published his book "Die Entstehung der Kontinente" (The Origin of Continents) describing a continental and an oceanic crust. At the time this was very speculative because Wegener was not a "geologist". But after 1960, the concept of plate tectonics with continental drift became generally accepted: The oceanic crust is upwelling along mid-oceanic ridges and spreading at a speed of ~2 cm/year. Continents are also moving at similar velocities over large distances. Around 300 My (Permian) continents formed a giant super-continent called Pangaea (Fig. 2.2). South America and Africa were joined and indeed rock formations and fauna are very similar in the east of South America and in the west of Africa. When Pangaea started splitting around 200 My ago (Triassic), a sea called Tethys formed at the equator between Laurasia in the north and Gondwanaland in the south. Sediments such as limestone and sandstones were deposited in this ocean on the

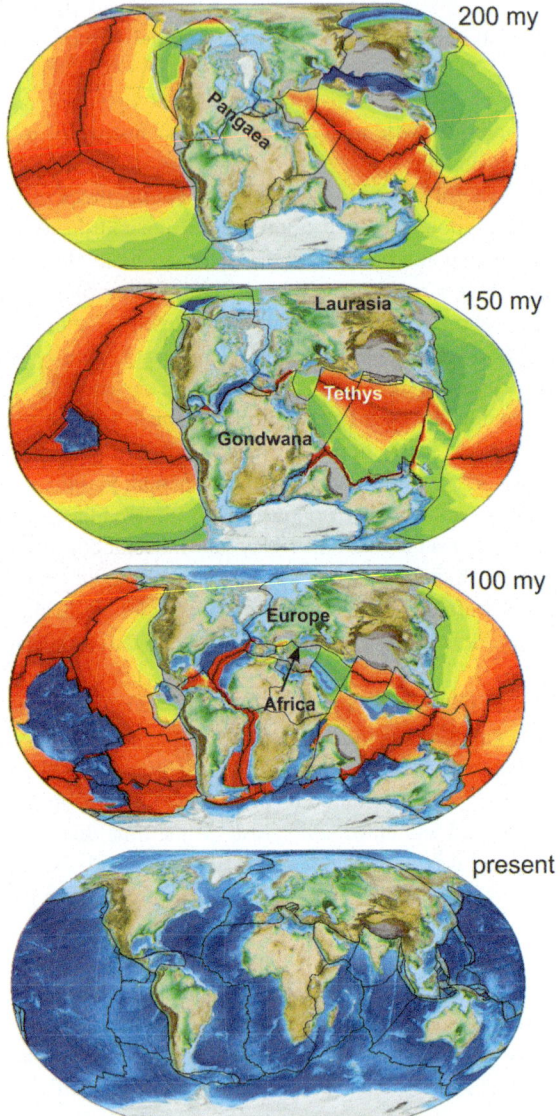

Fig. 2.2 Continental movements over 200 My. Modified from Seton et al. (2012). Colors on oceans indicate age of oceanic lithosphere (red is young, green is old)

edge of the continents. We can recognize these sediments in the Bergell after they have been converted into marble and quartzite. In the Cretaceous period (70 My) Laurasia collided with Gondwanaland. Parts of North Africa and Adriatic Tethys units were pushed north towards Europe and this collision led to the formation of the Alps.

The continental collision caused a stack of thrust sheets (nappes) with components of the European crust in the north (and at depth) and the African and Mediterranean crust (Apulian microcontinent) in the south (pushed over). In an intermediate expansion phase, basaltic magmas extruded along a mid-ocean ridge into the Tethys Sea. Even parts of the upper Earth's mantle—below the crust—with a peridotitic composition (predominantly olivine) came close to the surface. These oceanic units are the material from which rocks such as gabbro, amphibolite, prasinite (greenschist) and serpentinite transformed. Basalts were covered by sediments.

Figure 2.3a shows a picture of the various geological units in the Alps (Pfiffner 2016). The blue units are parts of the European continent (*Pennine*), the brown are parts of Africa (*Austroalpine*, also known as East-Alpine). Figure 2.3b is a NS cross-section through the Alps, just west of the Bergell.

Figure 2.4 displays the individual tectonic units in the Bergell Alps and you will have to become familiar with names of nappes. The deeper *Pennine nappes* in the northwest belonged to the Eurasian continent (Gruf, named after Monte Gruf, S of Villa di Chiavenna, Tambo, named after Pizzo Tambo W of Splügen Pass, Suretta, named after Surettahorn E of Splügen Pass, Platta, named after Piz Platta W of Marmorera, and Avers, named after Val Avers S of Andeer). They are blue in Fig. 2.3a.

The higher *Austroalpine nappes* are parts of the Apulian-African continent in the east and south (the lowest is Margna, named after Piz da la Margna, S of Maloja, above it Err, named after Piz d'Err SE of Savognin, Julier, named after Piz Julier NE of Julier Pass, and Bernina, named after Piz Bernina S of Pontresina). They are brown in Fig. 2.3a (e.g. Trümpy 1975). In the Maloja area of the Bergell the two tectonic domains are in direct contact. The Bergell granite intruded into the highest parts of the Pennine nappes (red in Figs. 2.3a and 2.4).

The border between the Upper Pennine and lower Austroalpine is particularly interesting. This *Platta* nappe and related mafic-ultramafic rocks in Valmalenco consist mainly of oceanic units (green in Fig. 2.4) and extends from Valmalenco in the south with a lot of serpentinite, along the Bergell granite with amphibolite, to Arosa (e.g. Bernoulli and Weissert 1985; Weissert and Bernoulli 1985). On some excursions we will explore this zone representing the Tethys Sea.

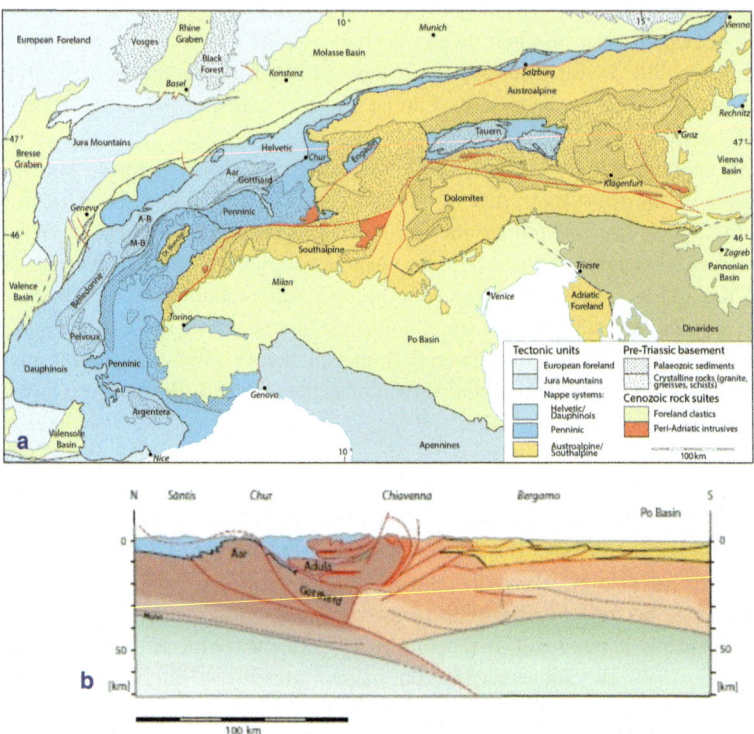

Fig. 2.3 a Tectonic structure of the Alps: blue are "European" units, including Pennine nappes, brown are "African" units, including Austroalpine nappes and Southalpine units. Beige are Tertiary sedimentary basins. Red are Tertiary intrusions. The red lines are faults. **b** NS cross-section across the Alps (courtesy A. Pfiffner)

The Bergell granite intruded into the stack of nappes in the Oligocene (Tertiary) and was described by Staub (1918). A profile of Staub (1958, Fig. 2.5a) displays it as a large intrusion that can be compared with the Hercynian (Carboniferous) Aar granite in the Central Alps or the Jurassic Sierra Nevada granite in California. However, subsequent field studies later revealed that the young Bergell granite is much more local and is closely associated with tectonic movements during the Alpine collision (Fig. 2.5b; Wenk 1973). Figure 2.6 gives a sketch of the geological-tectonic history of the Bergell Alps, from the Mesozoic over the Tertiary to the present.

2 The Geological Framework

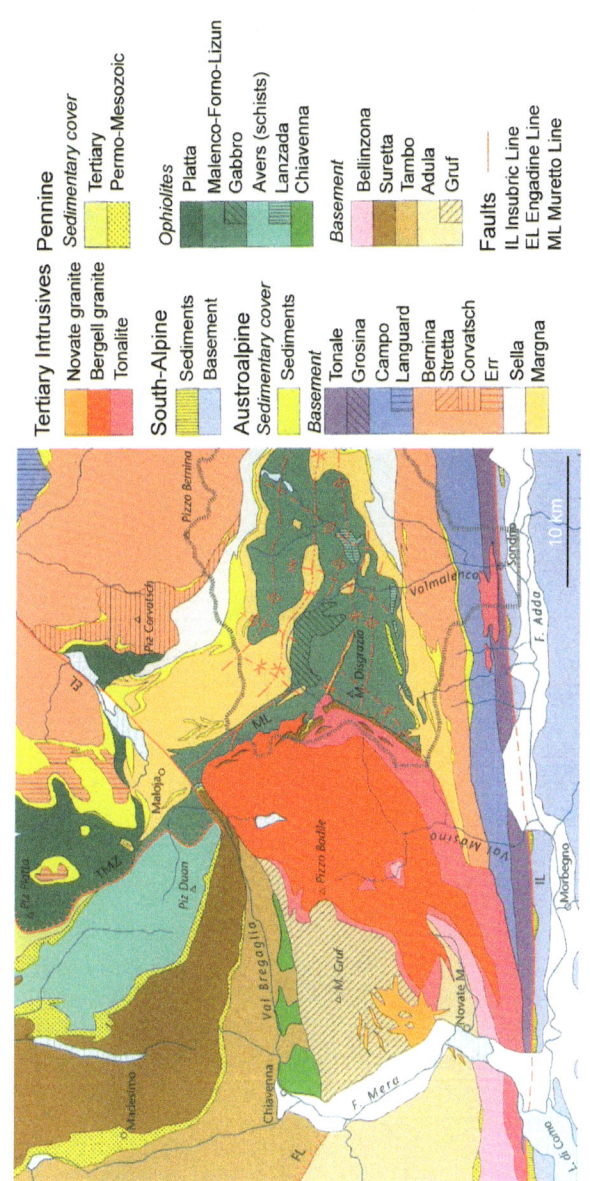

Fig. 2.4 Map of tectonic units in the Bergell Alps (courtesy A. Montrasio)

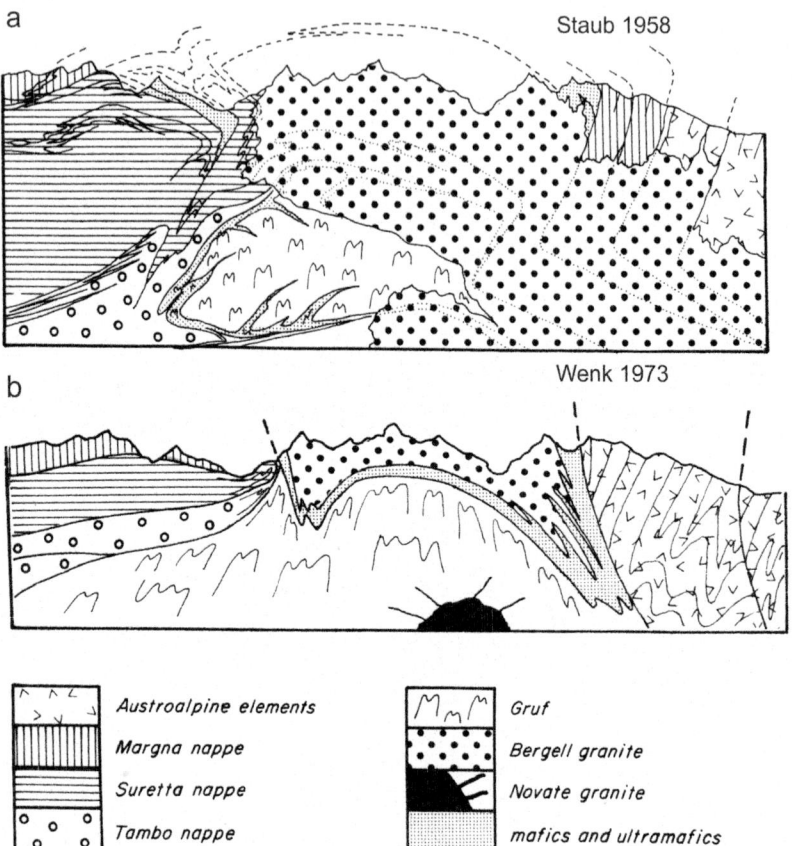

Fig. 2.5 Cross sections of the Bergell intrusion with different interpretations: **a** Staub (1958) and **b** Wenk (1973)

Many rocks of the Bergell Alps are metamorphic rocks that recrystallized at different pressure and temperature conditions. Old granites, such as from the Pennine Tambo and Suretta nappes, transformed to gneiss, basalt to amphibolite and greenschist, limestone to marble etc. Based on diagnostic mineral associations, rocks can be assigned to a metamorphic facies (e.g. Turner 1981). Figure 2.7

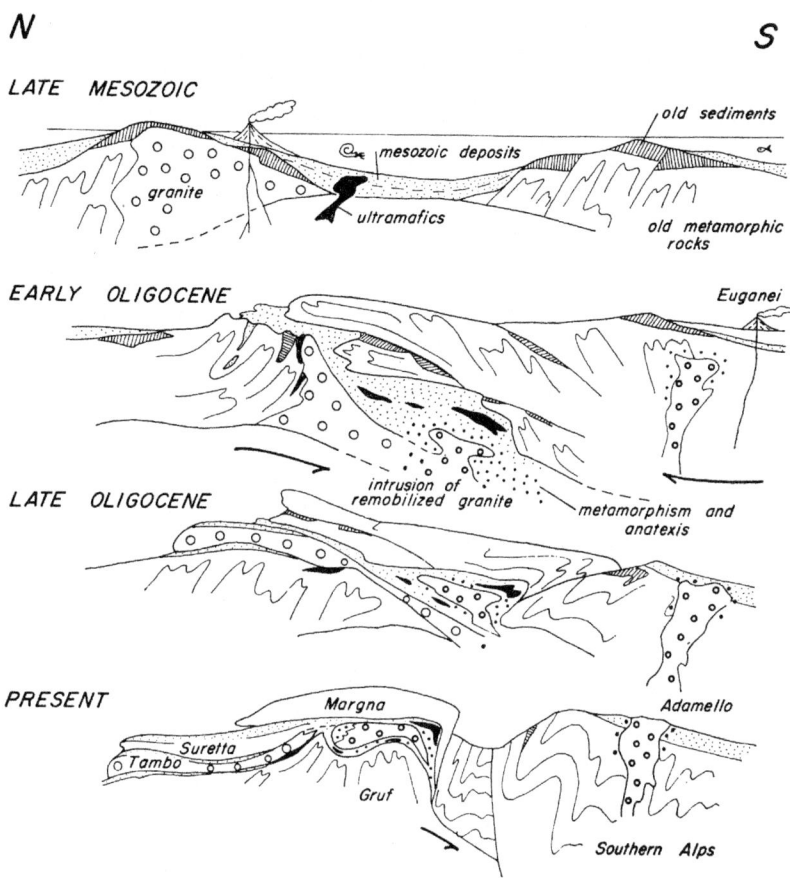

Fig. 2.6 Schematic sketch illustrating the geological history of the Bergell. In late Mesozoic old continents were separated by the Tethys Ocean. In early Oligocene they collided, initiating thrusts. In late Oligocene these tectonic movements came to an end and erosion started, producing present day morphology (after Wenk 1973)

shows a temperature–pressure (depth) diagram of typical facies. The minimum melting of granite as function of pressure and temperature is also indicated.

During the Alpine tectonic events in the Bergell, rocks of the Suretta, Platta, Err and Margna nappes recrystallized in *greenschist facies* at relatively low

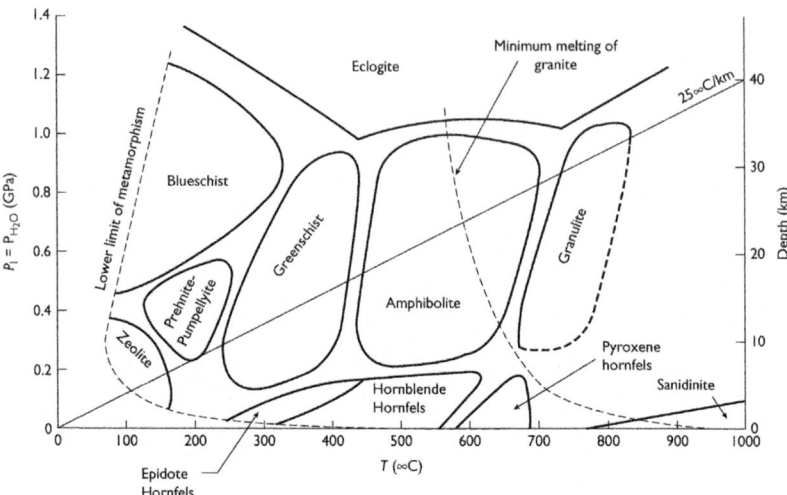

Fig. 2.7 Metamorphic facies as function of pressure/depth and temperature. The diagonal line is the average geothermal gradient in the continental crust (25 °C/km). The dashed line is the minimum melting of granite

temperature 250–450 °C and pressure 0.2–1 GPa as illustrated in a temperature–pressure phase diagram (Fig. 2.7). Rocks of the deeper Tambo nappe were buried about 15 km, with temperatures of 400–550 °C, typical of *amphibolite facies*. The high temperatures in the deeper western Pennine crust, such as the Gruf gneiss formed at *granulite facies* conditions with characteristic high-temperature mineral assemblages such as garnet-cordierite-sapphirine. At these high temperatures (~ 700 °C) there is local mobilization (migmatites), and finally melting and intrusion of Bergell granite, tonalite (~ 30–35 My) and Novate Granite (~ 20–25 My). Tonalite is a mafic granodiorite named after Passo Tonale near Sondrio IT. The intrusion of the granite into higher units, with continuous NS shortening, creates overthrusts and folding of the granite. The intrusion of relatively thin layers of melt is associated with rapid cooling so that the final phase of the emplacement still takes place plastically but largely in a solid state, which is evident in the regular planar fabric of the granite with oriented feldspar megacrysts, for example in Val Bondasca.

2 The Geological Framework

A word about geological profiles. These are often based on seismic measurements or deep drilling (for example in the Molasse basin of northern Switzerland, Fig. 2.3b) and partly hypothetical (Fig. 2.5a). But in the Bergell, with a 3-dimensional topography of high peaks and steep valleys, you can construct the profiles directly from the geological map and surface outcrops (e.g. Fig. 2.8).

Figure 1.2 showed a cross-section of the Maloja region from Studer's 1851 study. Figure 2.9 are profiles at a similar location by Cornelius (1950). In the latter map (after 100 years) we see many more details about the tectonics on the north side of the Bergell with Pennine nappes at depth (Suretta, Platta) and Austroalpine nappes above (Margna, Err, Bernina). Nappe overthrusts began in the late Cretaceous, but were mainly active in the Eocene and early Miocene. First, there was the overthrust of the Austroalpine, with an east–west shortening and large folds. An extension phase is followed by a new compression, this time north–south, in which the Austroalpine was thrust over the Pennine nappes.

On the north side of the Bergell we see the stack of Pennine nappes, Tambo at Soglio, above it Suretta at Piz Cam, and the Avers nappes at Piz Duan (Fig. 2.10a). They are parts of the European plate. In the south of the Upper Engadine we find the Austroalpine nappes Carungas, Sella, Corvatsch and Bernina with a view of Val Fex and behind it Piz Bernina with the Bianco-Grat, the highest mountain of Graubünden (4049 m) (Fig. 2.10b). They are parts of the old African continent. In the southwest of the Bergell we see the Bergell granite, here Val Albigna with Cima di Castello (Fig. 2.10c) which intruded into the upper part of the Pennine nappes.

The stack of Alpine nappes is separated from the Southern Alps, with a crystalline basement and sedimentary cover, by the Insubric/Tonale Line (red lines in Figs. 2.3a and 2.4). This thrust fault has a right-lateral displacement of >80 km (right-lateral means that looking at the fault, the opposite part has moved to the right), and the northern part is tectonically 10–20 km deeper. The town Novate, south of Chiavenna (N of the line) was once near Pallanza on Lago Maggiore (S) in the west, and the Adamello granite (on the South side of the Insubric Line) was much further east. The Insubric Line displacement developed after the nappe overthrusts and the granitic intrusions.

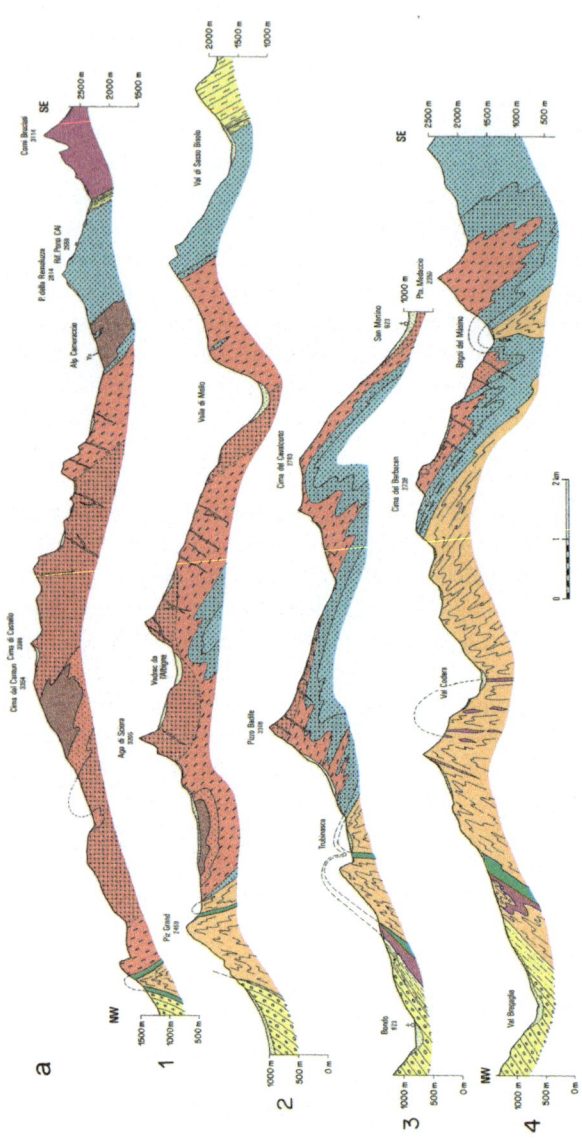

Fig. 2.8 a N–S cross-sections through the Bergell granite with complex deformation features. **b** Legend of geological units. **c** Traces of cross sections (modified after Wenk 1992)

2 The Geological Framework

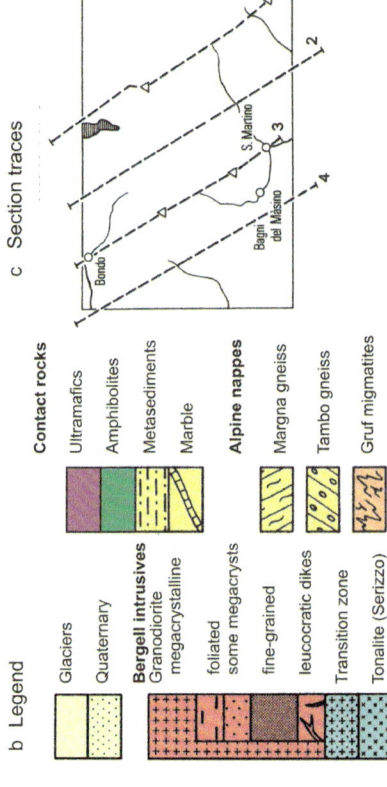

Fig. 2.8 (continued)

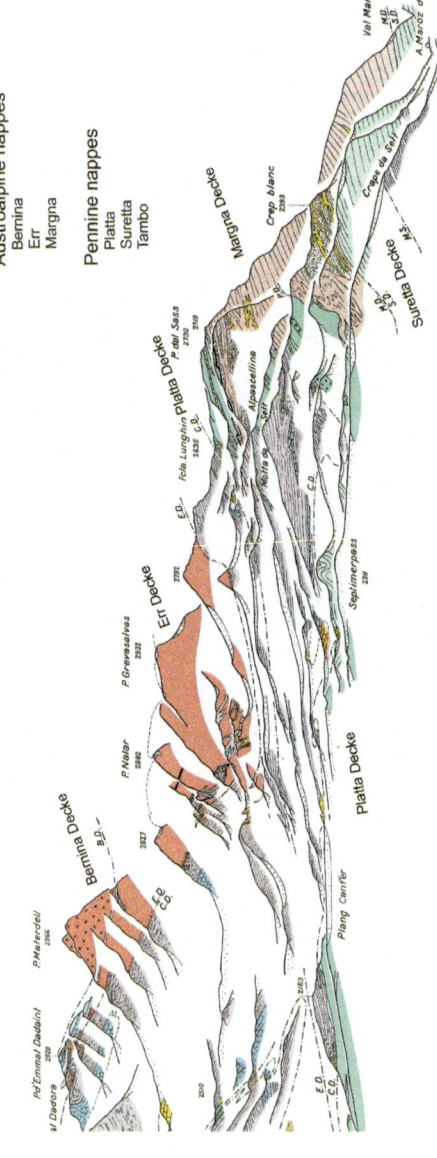

Fig. 2.9 NS cross sections with nappes on the N-side of the Bergell/Engadine (modified from Cornelius 1950)

Fig. 2.10 Views of the landscape. **a** N-side of Val Bregaglia with Pennine nappes Tambo, Suretta and Avers. This is old "Europe". **b** View across Lake Sils with Austroalpine nappes Carungas, Sella, Corvatsch, Bernina (Err). This is part of "Africa". **c** Southside of Val Bregaglia in Val Albigna with young Bergell granite. On the left is Cima di Castello

References

Bernoulli D, Weissert H (1985) Sedimentary fabrics in Alpine ophicalcites, South-Pennine Arosa zone, Switzerland. Geology 13:755–758

Burton-Johnson A, Macpherson CG, Ottley CJ, Nowell GM, Boyce AJ (2019) Generation of the Mt. Kinabalu granite by crystal contamination of intraplate magma modelled by equilibrated major element assimilation with fractional crystallization (EME-AFC). J Petrol 60(7):1461–1487

Cornelius HP (1950) Geologie der Err-Julier-Gruppe. II. Teil: Der Gebirgsbau. Beitr Geol Karte Schweiz [N.F.] 70(2):264

Galli A, Le Bayon B, Schmidt MW, Burg J-P, Reusser E, Sergeev SA, Larionov A (2012) U–Pb zircon dating of the Gruf Complex: disclosing the late Variscan granulitic lower crust of Europe stranded in the Central Alps. Contrib Mineral Petrol 163:353–378

Gnägi C, Labhart TP (2017) Geologie der Schweiz, 10th edn. Ott Verlag, 208 pp

Grünenfelder M, Stern TW (1960) Das Zirkon-Alter des Bergeller Granits. Schweiz Mineral Petrogr Mitt 40:253–259

Kamleitner S, Ivy-Ochs S, Monegato G, Gianotti F, Akcar N, Vockenhuber C, Christl M, Synal H-A (2022) The Ticino-Toce glacier system (Swiss-Italian Alps) in the framework of the Alpine Last Glacial Maximum. Quat Sci Rev 279:107400

Liati A, Gebauer D, Fanning M (2000) U–Pb SHRIMP dating of zircon from the Novate granite (Bergell, Central Alps): evidence for Oligocene-Miocene magmatism, Jurassic/Cretaceous continental rifting and opening of the Valais trough. Schweiz Mineral Petrogr Mitt 80:305–316

Pfiffner OA (2016) Basement-involved thin-skinned and thick-skinned tectonics in the Alps. Geol Mag 153(5/6):1085–1109

Pfiffner OA (2019) Landschaften und Geologie der Schweiz. Haupt Verlag, 360 pp

Seton M, Mueller RD, Zahirovic S, Gaina C, Torsvik T, Shephard G, Talsma A, Gurnis M, Turner M, Maus S, Chandler M (2012). Earth-Science Rev 113:212–270

Staub R (1918) Geologische Beobachtungen am Bergellermassiv. Vjschr natf Ges Zürich 63:1–18

Staub R (1958) Klippendecke und Zentralalpenbau. Beziehungen und Probleme. Beitr Geol Karte Schweiz [N.F.] 103:184

Trümpy R (1975) Penninic-Austroalpine boundary in the Swiss Alps: a presumed former continental margin and its problems. Am J Sci 275A:209–238

Turner FJ (1981) Metamorphic petrology. Mineralogical, field and tectonic aspects, 2nd edn. McGraw-Hill, 524 pp

Valley JW, Cavosie AJ, Ushikobo T, Reinhard DA, Lawrence DF, Larson DJ, Clifton PH, Kelly TF, Wilde SA, Moser DE, Spicuzza MJ (2014) Hadean age for a post-magma-ocean zircon confirmed by atom-probe tomography. Nat Geosci 7:219–223

von Quadt A, Grünenfelder M, Büchi H (1994) U–Pb zircon ages from igneous rocks of the Bernina nappe system (Grisons, Switzerland). Schweiz Mineral Petrogr Mitt 74:373–382

Wegener A (1915) Die Entstehung der Kontinente und Ozeane, 1st edn. Viehweg, 231 pp

Weissert H, Bernoulli D (1985) A transform margin in the Mesozoic Tethys: evidence from the Swiss Alps. Geol Rundsch 74:665–679

Wenk H-R (1973) The structure of the Bergell Alps. Eclogae Geol Helv 66(2):255–291

Wenk H-R (1992) Foglio 1296 Sciora, Atlante geologico della Svizzera 1:25000, Note esplicative. Servizio idrologico e geologico nazionale della Svizzera, pp 61

Six Highlights of Bergell Geology

3

In this section we give some background about minerals and rocks of the Bergell, discuss Bergell granite in some detail and provide short summaries on important issues such as microstructures and formation conditions. We review the recent geological history with glaciers and landslides. Then topics such as mining, rock quarries and water exploitation are described. These issues should be reviewed before starting an excursion. You may also want to follow up with more detailed information in books about mineralogy, petrology and geology.

3.1 Minerals and Rocks in the Bergell

Rocks are composed of minerals. We first describe typical minerals of the Bergell, some very common ones and also some rarities, like chiavennite named after the town Chiavenna at the foot of Val Bregaglia in Italy.

Minerals are chemically homogeneous compounds with a regular crystal structure and periodic arrangements of atoms caused by bonding forces between atoms. The most common minerals in the Earth's crust are silicates with $[SiO_4]$ tetrahedra linked to form chains, sheets or frameworks. The periodic crystal structure often leads to regular crystal forms such as cubes for pyrite and octahedra for magnetite. Tables 3.1, 3.2 and Figs. 3.1 and 3.2 give some examples of Bergell minerals, their chemical composition and characteristic properties. Where minerals make up more than 5% of a rock, they are called rock-forming minerals such as feldspar, quartz and mica in granite, hornblende and plagioclase in amphibolite and tonalite, quartz in quartzite, calcite or dolomite in marble.

Table 3.1 Bergell minerals: sulfides, oxides, carbonates and sulfates

Name	Composition	Color	Hardness (Mohs)	Cleavage	Typical properties	Figures
Sulfides						
Pyrite	FeS_2	Yellow/metallic	6	None	Cube with striations	3.1a
Chalcopyrite	$CuFeS_2$	Yellow-greenish-red	4	None	No striations	3.21a
Oxides						
Quartz	SiO_2	No color transparent	7	None	Prismatic, no shape in rocks	3.1b
Magnetite	Fe_3O_4	Black/metallic	6	Octahedral	Oct.	4.4e
Carbonates						
Calcite	$CaCO_3$	White	3	Rhombohedral	HCl fiz	3.1c
Dolomite	$CaMg(CO_3)_2$	White	4	Rhombohedral	HCl fiz (conc.)	
Kutnahorite	$CaMn(CO_3)_2$	Pink	3.5			
Azurite	$Cu_3(CO_3)_2(OH)_2$	Blue	3.5			
Malachite	$Cu_2(CO_3)(OH)_2$	Green	3.5			3.1d, 3.21b
Sulfates						
Gypsum	$CaSO_4 \cdot 2H_2O$	White	1.5	Excellent	Water soluble	3.1e

Table 3.2 Silicate minerals from the Bergell Alps

Name	Composition	Color	Hardness	Cleavage	Typical	Figures
Orthosilicates						
Garnet	$Fe^{2+}_3 Al_2(SiO_4)_3$	Red	7	Equiaxed		3.1f, 4.19b
Olivine	$(Mg,Fe)SiO_4$	Green	7			
Clinohumite	$Mg_9(SiO_4)_4(OH)_2$	Yellow	6.5			
Epidote	$Ca_2Fe^{3+}Al_2SiO_4(Si_2O_7)O(OH)_2$	Green	6.5		Prismatic	3.1g
Andalusite	Al_2SiO_5	Pink	7.5		Prismatic	3.11b
Sillimanite	Al_2SiO_5	White	6.5	Fibers		3.10b
Kyanite	Al_2SiO_5	Blue	6.5		Prismatic	3.11b
Staurolite	$(Fe^{2+},Mg)_2(Al,Fe)_9O_6(SiO_4)O(OH)$	Brown	6.5		Prismatic	
Beryl	$Be_3Al_2Si_6O_{18}$	Blue/green	7.5		Prismatic	3.1h
Zircon	$Zr[SiO_4]$	Brown	7.5			3.1j, 4.23g
Cordierite	$Al_3(Mg,Fe^{2+})_2[Si_5AlO_{18}]$	Blue	7			4.23e
Titanite	$CaTi[SiO_5]$	Brown	5			3.1k
Tourmaline	$NaFe^{2+}_3(AlSiO_3)_6(BO_3)_3(OH)_4$	Black	7		Prismatic	3.11
Chloritoid	$Fe^{2+}AlAl_3(SiO_4)_2O_2(OH)_4$	Green	6.5	Platy		4.15b–d
Sapphirine	$Mg_2Al_4[O_6/SiO_4]$	Blue	6			4.23f
Chain silicates						
Pyroxenes						
Diopside	$CaMg[Si_2O_6]$	Light-green	6	$\{110\}\ 87°$		

(continued)

Table 3.2 (continued)

Name	Composition	Color	Hardness	Cleavage	Typical	Figures
Amphiboles						
Tremolite	$Ca_2Mg_5[Si_8O_{22}](OH)_2$	White	5.5	{110} 124°		3.2a
Actinolite	$Ca_2Fe^{2+}_5[Si_8O_{22}](OH)_2$	Green	5.5	{110} 124°		
Hornblende	$(Na,K)_{0-1}(Ca,Na)_2(Mg,Fe^{2+})$ $(Al,Fe^{3+})Si_7AlO_{22}(OH,F)_2$	Green	5.5	{110} 124°		
Riebeckite	$Na_2Fe^{2+}_3Fe^{3+}_2[(OH)/Si_4O_{11}]_2$	Dark blue	5.5	{110} 124°		4.13b-c
Others						
Rhodonite	$Ca_4Mn[Si_5O_{15}]$	Pink/red	6	Good		
Wollastonite	$Ca_3[Si_3O_9]$	White	4.5	Excellent	Fibers	
Sheet silicates						
Muscovite	$KAl_2[Si_3Al_2O_{10}](OH)_2$	Clear-white	2.5			3.2b
Mariposite	$K(Al,Cr)_2[Si_3AlO_{10}](OH)_2$	Green	2.5	(001)	Platy	3.2c
Biotite	$K(Mg,Fe)_3[Si_3AlO_{10}](OH)_2$	Brown	2.5	(001)	Platy	3.2d
Chlorite	$Mg_3(OH)_6Mg_2Al(OH)_2[Si_3AlO_{10}]$	Green	2.5	(001)	Platy	3.2e
Talc	$Mg_3[Si_4O_{10}](OH)_2$	Light green	1	(001)	Platy	
Antigorite	$Mg_3[Si_2O_5](OH)_4$	Dark green	3	(001)	Platy	
Lizardite	$Mg_3[Si_2O_5](OH)_4$	Green	3	(001)	Platy	3.2f
Chrysotile	$Mg_3[Si_2O_5](OH)_4$	White	2.5	(001)	Fibrous	3.2g
Stilpnomelane	$KFe_8[Si_{11}AlO_{28}](OH)_8 \cdot 2H_2O$	Black	3.5	(001)		3.10d

(continued)

Table 3.2 (continued)

Name	Composition	Color	Hardness	Cleavage	Typical	Figures
Framework silicates						
Feldspars						
Microcline	K[AlSi$_3$O$_8$]	White	6	(001)(010)		3.2h
Orthoclase	K[AlSi$_3$O$_8$]	White	6	(001)(010)		3.6
Albite	Na[AlSi$_3$O$_8$]	White	6	(001)(010)		3.10c
Anorthite	Ca[Al$_2$Si$_2$O$_8$]	White	6	(001)(010)		3.10c
Plagioclase	(Na,Ca)[(Si,Al)Si$_2$O$_8$]	White	6	(001)(010)		3.10a
Zeolites						
Laumontite	Ca[Al$_2$Si$_4$O$_{12}$]·4H$_2$O	White	3	(010)(110)		3.2i
Chiavennite	CaMnBe$_2$Si$_5$O$_{13}$(OH)$_2$·2H$_2$O	Pale orange	3			3.1i

Fig. 3.1 Minerals: **a** Pyrite, Val Febbraro, 5 mm. **b** Quartz, Starleggia, up to 10 cm. **c** Calcite, Val d'Oro, 5 cm. **d** Malachite Piz Mungiroi. **e** Gypsum, Cavi, 15 cm. **f** Garnet (spessartine), Sivigia, 1 cm. **g** Epidote, Preda Rossa, 31 mm long. **h** Beryl, Sivigia, 15 mm long. **i** Chiavennite, Tanno, 3 mm. **j** Zircon, La Tieda/Montaccia, 1.5 mm. **k** Titanite, Tubladel, 12 mm long. **l** Tourmaline, Cma. Di Codera, 30 mm long (all, except **d**, **e** courtesy of Bedogné et al. 1995)

It is rare to find minerals with regular shapes and flat surfaces as illustrated in Figs. 3.1 and 3.2 in rocks. They mostly occur in open fissures. There are exceptions in pegmatite dikes or in some contact rocks adjacent to igneous intrusions. Garnet, epidote and beryl are examples (Fig. 3.1f–h). The exhibits in the valley museum Ciäsa Granda in Stampa and the Grazioli collection in the Museo dei Minerali in Sondrio provide excellent displays.

3.1 Minerals and Rocks in the Bergell

Fig. 3.2 Minerals. **a** Fibrous tremolite vein in serpentinite, Tavretga/Bivio, 2 cm wide. **b** Muscovite, 2 cm wide. **c** Mariposite, Avers, 1 cm wide. **d** Biotite, sample 5 cm wide. **e** Clinochlore, quarry Piuro, sample 5 cm wide. **f** Lizardite, sample 10 cm wide. **g** Fibrous chrysotile, Uschione, 40 cm wide. **h** Bergell granite with large alkali feldspar, smaller (white) plagioclase, some clear quartz and black biotite, Albigna, sample 7 cm wide. **i** Laumontite, Stoveno, crystals up to 14 mm (**b**, **d** courtesy of W. Hunkeler, **g**, **i** Bedogné et al. 1995)

How do we identify minerals? One characteristic is hardness and is described with the Mohs hardness scale ranging from 1 to 10. Talc (hardness 1) can be scratched with a fingernail (2.5), also gypsum (1.5). Calcite (hardness 3) can be scratched with a steel nail or a pocket knife blade (6). One of the softest minerals is talc, the hardest (10) is diamond.

Some minerals such as mica, feldspar and calcite have a regular fissile cleavage on one or several planes at characteristic angles. Sheet silicates such as micas and chlorite have a single excellent cleavage, feldspars have two cleavages, roughly at right angles and calcite has three cleavages on rhombohedral planes. Hornblende has two cleavages at roughly 120° angles. This can be examined with a hand lens. Quartz has no cleavage and fractures on irregular surfaces like glass.

Colors vary widely: Feldspars are mostly white, hornblende and actinolite are dark green, garnets are red, kyanite is blue, malachite is green. Figures 3.1 and 3.2 give an overview.

Crystal shapes are complicated and depend on the crystal structure. There is the familiar picture of quartz with a prism capped by a pyramid (Fig. 3.1b) but most quartz in rocks has no particular shape.

For mineral identification you need to rely on a corresponding guide (e.g. Garlick 2014; Meyer 2017; Weibel et al. 1990). Three reviews of Bergell minerals are Maurizio and Weibel (1982), Maurizio and Meisser (1993) and Bedogné et al. (1995). So far 209 different mineral species have been identified.

Some rare minerals of the Bergell Alps are orange chiavennite $CaMnBe_2Si_5O_{13}(OH)_2 \cdot 2H_2O$ (Fig. 3.1i), transformed from beryl, and first discovered in a pegmatite vein above Tanno (S-Chiavenna) (Bondi et al. 1983), pink kutnahorite $(Mg_{0.2}Mn_{0.2}Fe_{0.1}Ca_{0.5})CO_3$ occurring at Piz Cam (767.1/137.7), originally found in Czechoslovakia (Wenk and Maurizio 1978), and the alumosilicate mullite $\sim 3Al_2O_3 \cdot 2SiO_2$ in the contact zone of tonalite with schists in Val Schiesone/Malenco (Wenk 1983). An unusual chromium-containing mica is bright green mariposite (also known as fuchsite) in schists of the upper Pennine nappes (Fig. 1.2c). Many of the minerals in these tables and figures will be referred to in the following chapters and fieldtrip itineraries.

Rocks are divided into sedimentary rocks, igneous rocks and metamorphic rocks. *Sedimentary rocks* were deposited on the surface of the Earth. Sediments such as sand, gravel, clay and organic fragments compacted to form limestone, sandstone and shales. *Igneous rocks* originated from a melt: *volcanic rocks* were molten at depth, then the melt extruded to the surface and solidified (such as basalt), *plutonic rocks* also began as melt at depth and then intruded the continental crust, slowly solidifying (a typical plutonic rock is granite). *Metamorphic rocks* where parent rocks—sedimentary rocks, volcanic and plutonic rocks, or older metamorphic rocks—recrystallized at different pressures and temperature conditions.

In the Bergell region there are only plutonic and metamorphic rocks and some sediments such as moraines, river alluvium, talus debris and landslides. Older sedimentary rocks, volcanic and plutonic rocks were transformed during the Alpine metamorphism and recrystallized with new minerals growing. However, we will observe on some excursions features of the original rocks that have been preserved such as pillow structures typical of oceanic basalt, in amphibolite at Muretto Pass and fossil radiolaria skeletons in siliceous slates above Grevasalvas.

3.1 Minerals and Rocks in the Bergell

Plutonic rocks crystallized from a melt (magma) at depth. The most important are rocks of granitic composition with quartz, plagioclase and alkali feldspar as dominant minerals and also some muscovite and biotite. Old granites (e.g. Julier) reach the Bergell valley in moraines. In these ancient granites (~ 300 My), mica and also plagioclase are often partially converted into chlorite and illite (a clay mineral).

Among the young granites we will explore the *Bergell granite* (30–35 My) on five excursions (Fig. 3.3a) and the youngest Novate granite (20–25 My) on one excursion (4.2.11, Fig. 3.3b). Their mineralogical composition will be discussed in detail in Sect. 3.2. Much of Bergell granite consists of large isometric alkali feldspar crystals (megacrysts), fine-grained white plagioclase and transparent quartz. Dark spots of biotite are easy to recognize (Fig. 3.3a). *Novate granite* (in Italy also known as San Fedelino granite) is more homogeneous, with less biotite (Fig. 3.3b). Another variety of the Bergell intrusion consists of *tonalite* (Fig. 3.3c), named after Passo Tonale. In Italy tonalite is also known as "serizzo". It is darker than granite with dark green hornblende as a major component, along with plagioclase and some quartz. It surrounds the Bergell granite like a shell (blue-green in Fig. 2.8a) and dominates in Val Màsino. In Valmalenco tonalite is in contact with amphibolite and in the contact zone fragments of amphibolite are often enclosed in tonalite (Fig. 3.3c). Small intrusions of tonalite can be observed just N of the Insubric Line (e.g. Triangia, N of Sondrio in lower Valmalenco).

Obviously, temperatures were high enough to melt granitic and tonalitic compositions, but on the west side of the Bergell Alps older rocks of granitic composition were only partially mobilized. These so-called *migmatites* of the Gruf zone were often compositionally segmented and intensely folded (Fig. 3.3d).

Dikes intruded as melt or hydrothermal solutions along fractures. In Bergell granite, dikes of silica-rich aplite and pegmatite are common (Fig. 3.3e). The composition of these dikes has less plagioclase and biotite than granite. They are residual melts with a higher water content (hydrothermal) and therefore a lower melting point. *Aplite* (large dike) is homogeneous and relatively fine-grained. *Pegmatite* is more coarse-grained and often zoned. Pegmatites, e.g. in Val Albigna and Val Bondasca, contain secondary minerals such as beryl and garnet. In the contact zone there are also veins and dikes of quartz, some containing beryl. These quartz dikes were not formed from melts but crystallized from concentrated hydrous solutions where the remaining SiO_2 content precipitated.

Less common are mafic dikes, mainly composed of plagioclase, pyroxene and amphibole and a dark green color, so-called *diabase*. There are old diabase dikes that intruded into the Hercynian granites at Piz Grevasalvas and Piz Lagrev (Fig. 3.3f). Here many mineral components were transformed during subsequent

Fig. 3.3 Igneous rocks in the Bergell Alps. **a** Bergell granite (Albigna). **b** Novate granite (Novate-Mezzola). **c** Tonalite with xenoliths (Val Sissone). **d** Migmatite (Val Codera). **e** Aplite and pegmatite dikes in granite (Val Forno). **f** Mafic dike in Err granite (Piz Materdell). (Hammer is 50 cm long, pocket knife 8 cm)

Alpine metamorphism. Then there are some young diabase dikes of similar age as the Bergell granite (35–40 My), for example at Piz Lizun and Alpe Furcela (excursion 4.2.5).

Metamorphic rocks are important in the contact zone of the granite but also in the stack of nappes on the north and east sides of the Bergell.

3.1 Minerals and Rocks in the Bergell

Dominant among metamorphic rocks is *gneiss* with a layered texture formed mainly by recrystallization under stress and strain. The main mineral components are feldspar, quartz and mica. *Orthogneiss* transformed from ancient granite, *paragneiss* from sedimentary rocks such as clay and shale. As deformation increases, gneiss transforms into schists.

A unique granitic gneiss is Tambo gneiss (Fig. 3.4a), famous for its exceptional cleavage. On the cleavage surface you can recognize a lineation top left to bottom right. This rock is mined in quarries at Soglio and Promontogno for plate production (tables, roof tiles etc.) (see Sect. 3.6.3).

Temperature and pressure conditions can be estimated from the mineralogical composition of metamorphic rocks (Fig. 2.7). Chlorite gneiss, so-called greenschist, crystallized at low temperature (higher nappes), biotite schists with hornblende represent higher temperatures. We will examine this in Sect. 3.4.

Greenschists, light-green in color and consisting of chlorite, muscovite, quartz and albite, is widespread in the higher Pennine nappes (Suretta, Avers) (Fig. 3.4b). It is of similar chemical composition as amphibolite and transformed from mafic igneous rocks such as gabbro and basalt.

Amphibolite, dark green, consists mainly of hornblende and plagioclase, sometimes with biotite and rarely garnet (Fig. 3.4c). Also this rock was originally basalt deposited in the Tethys Ocean. In the Bergell and Val Malenco basalt was transformed into amphibolite at high temperature (Fig. 2.7, "Amphibolite facies"). In the Forno area we find amphibolite in contact with Bergell granite, often with aplite and pegmatite dikes (the older dike is aplite which was later crosscut by the larger pegmatite dike with zonation and red garnet) (Fig. 3.4c).

Serpentinite (Fig. 3.4d) is an Mg–Fe rich and SiO_2-poor rock (geologists call it ultramafic), consisting mainly of the mineral serpentine (in the Bergell all three polymorphs antigorite, lizardite Fig. 3.2f, and occasionally chrysotile, Fig. 3.2g, are observed). Polymorphs are minerals of the same composition but different crystal structures. The name "serpentine" comes from Latin "serpentinus": snake. Serpentine has high heat conduction and becomes warm in the sun, compared to insulating granite, and is therefore a preferred stone for reptiles (Lizardite is named after "lizard"). Serpentinite was originally mantle peridotite (an olivine rock: Mg_2SiO_4) which reacted with water when it was thrust into the upper crust (serpentine $Mg_3[Si_2O_5](OH)_4$). Olivine (an orthosilicate) transformed into serpentine (a sheet silicate).

Marble (Fig. 3.4e) was originally limestone, in the Bergell mainly calcite but sometimes dolomite as in marbles east of Casaccia (771.56/140.03). Compared to silicates, the carbonates can be deformed much more easily, and one finds local structures with intensive folding such as marble at Lägh da la Duana (Fig. 3.4e). In the field these carbonate rocks are easy to recognize, because with a pocket

Fig. 3.4 Metamorphic rocks in the Bergell Alps. **a** Tambo gneiss (Promontogno). **b** Greenschist (Val Maroz). **c** Amphibolite with pegmatite and aplite dike (Val Forno). **d** Serpentinite (Pass Lunghin). **e** Folded marble (Lägh da la Duana). **f** Quartzite (Val Maroz). (**a**, **b**, **f** are 12 cm wide, **c** is 30 cm wide, Hammer on **d** and **e** is 50 cm long)

knife or a hammer they can be scratched. Also, with diluted hydrochloric acid (HCl) calcite marble fizzes (dolomite marble fiz with concentrated HCl).

3.1 Minerals and Rocks in the Bergell

Quartzite (Fig. 3.4f), originally deposited as sandstone, was also a sedimentary rock and in the Bergell quartzites as well as most marbles are of similar Triassic age (~ 220 My). Marbles and quartzites mark tectonic boundaries and are easy to observe in the field (white bands). Another much rarer meta-sedimentary quartz-rich rock is redish radiolarite, e.g. at Grevasalvas and we will discuss it later (excursion 4.2.3).

Sediments As mentioned above, there are no "sedimentary rocks" but there are plenty of sediments that have been deposited on the surface, without being transformed or compressed into sedimentary rocks. These sediments such as moraines, river alluvium, talus on slopes under cliffs and glaciers are young (< 1 million years) but cover a large part of the surface of the Bergell Alps (~ 60%).

Talus debris can be found on most steep slopes below cliffs (Fig. 3.5a). It consists of angular fragments of rocks. The composition corresponds to that of the rocks exposed above. Rock falls from cliffs occur all year round but are dominant after heavy rainfalls and in spring when ice in fractures expands and melts.

Alluvium with rounded blocks of different sizes as well as sand can be found in rivers such as the Orlegna near Orden, the Maira with a large deposit S of Casaccia, above Vicosoprano and near Bondo. The composition of the boulders provides an insight into the geology of the catchment area (Fig. 3.5b and excursion 4.1.5).

Clay is a compact, very fine-grained aggregate of clay minerals (e.g. illite, montmorillonite) and is common in lakes such as Lake Albigna below the glacier (Fig. 3.5c) and Lake Sils.

Peat bog consists mainly of humus and developed in slight depressions of a horizontal postglacial surface by decomposition of trees (Fig. 3.5d). Deeper erosion created lakes such as Lägh da Cavloc and, of course, the much larger Lake Sils. In these organic and moist soils, an extraordinary flora developed, e.g. with carnivorous plants Drosera (bottom right of Fig. 3.5d) and Pinguicula (top right). At Maloja in Aira da la Palza, above the dam at Orden (773.5/140.2) and along the road to Cavloc (774.2/139.9) we can find excellent samples of these special plants. Over longer periods, peat bog transforms into lignite and, at higher temperatures, into hard coal, but not in the Bergell.

Moraines from glaciers mark the original extension of the glacier. Contrary to alluvium, debris in moraines is less rounded and contains more fine-grained clay-rich material and forms soils rich in nutrients (Fig. 3.5e). Vegetation brings humus that then mixes with sand and clay minerals. Thus, moraine surfaces are ideal for pastures and forests, relatively flat and fertile. There are large areas of the old valley glacier moraines (e.g. Grevasalvas/Blaunca, Palza/Maloja, Roticcio, Durbegia, Soglio/Tombal/Plan Vest) and also small moraines of secondary

Fig. 3.5 Sediments. **a** Talus slope underneath Monte del Forno. **b** River alluvium at Pranzaira. **c** Clay (illite) Val Albigna. **d** Peat bog near Orden/Maloja with Pinguicula (top) and Drosera (bottom). **e** Moraines on both sides of Forno glacier. **f** Ice with crevasse in Bondasca glacier

side glaciers without Bergell granite fragments (e.g. Motta Salacina and Ciresc-Vöga). In the main moraine near Tombal above Soglio, you can find pieces of Err granodiorite indicating that the glacier in the upper Engadine originally moved to the southwest.

Ice is another sediment transformed from snow in glaciers (Fig. 3.5f). Ice H_2O is also a mineral.

3.2 Bergell Granite

Granites intruded into the continental crust at depth and then became exposed on the surface by erosion. Most granites in Central Europe have a Variscan age (Carboniferous, ~300 My), such as Bernina, Err, Aar, Gotthard, Mt. Blanc, Vosges and Black Forest granites. Granite gneiss in the Bergell nappes, such as Tambo, Suretta and Margna, also originally intruded as granites in Carboniferous and were then transformed during the Alpine tectonic events in Cretaceous and Oligocene. Granites in the Bergell, Novate and Adamello are an exception and are much younger. They were mobilized during the collision of the African continent with Europe and intruded around 30 My (Grünenfelder and Stern 1960). In contrast to the old European granites these young granites are rare and much smaller by volume.

Typical Bergell granite is easy to recognize in hand specimens: Large (megacrystic) alkali feldspar crystals occur in a matrix of quartz and plagioclase, with dark biotite mica (Figs. 3.3a and 3.6). With a hand lens you can see quartz (transparent-glassy) and plagioclase (white). There are variants of Bergell granite where the large alkali feldspar megacrystals are missing.

Granitic rocks can be categorized based on volume ratios of the three main mineral components **A**lkali feldspar, **P**lagioclase and **Q**uartz, in a ternary APQ diagram (Fig. 3.7). In contrast to many "old" granites, Bergell granite is mainly "granodiorite", with more plagioclase P than alkali feldspar A. A second variety, tonalite, also formed from a melt, can be found below and above the Bergell granite (profiles in Fig. 2.8a), mainly on the south-side. Tonalite (Fig. 3.3c) contains hardly any alkali feldspar and is formed by mobilization of hornblende-rich rocks such as amphibolite.

The melting point of granitic rocks varies greatly with the water content and can drop to 700 °C as illustrated in a alkali feldspar (albite-orthoclase) phase diagrams (Fig. 3.8a–c, see also Fig. 2.7). After granite crystallizes, the water is enriched in the residual melt and later precipitates as a mixture of pegmatite and aplite, which cross the granite and contact rocks along fractures. Residual water causes chemical reactions in contact rocks, so-called hydrothermal

Fig. 3.6 Hand specimen of Bergell granite with large alkali feldspar crystals in a matrix of plagioclase, quartz and biotite (width is 10 cm)

metamorphism, often with large crystals of feldspar, quartz, beryl, garnet and tourmaline.

Apart from stream debris in the Maira (excursion 4.1.5) and in a few moraines, Bergell granite in Val Bregaglia is nowhere to be found on a road. The easiest way to reach it is by cable car to Albigna (excursions 4.1.4 and 4.2.6), in Forno Valley (excursion 4.2.1) or in the back regions of Val Bondasca (excursion 4.2.9). Figure 3.9a is a view into Val Bondasca from Tombal in autumn, with a moraine ridge in the foreground. Whatever is white here (snow covered) is Bergell granite. Everything else is underlying bedrocks such as Gruf migmatites and Tambo gneiss. Figure 3.9b is a view of Pizzo Cengalo and Pizzo Badile: all the rocks here are Bergell granite.

3.2 Bergell Granite

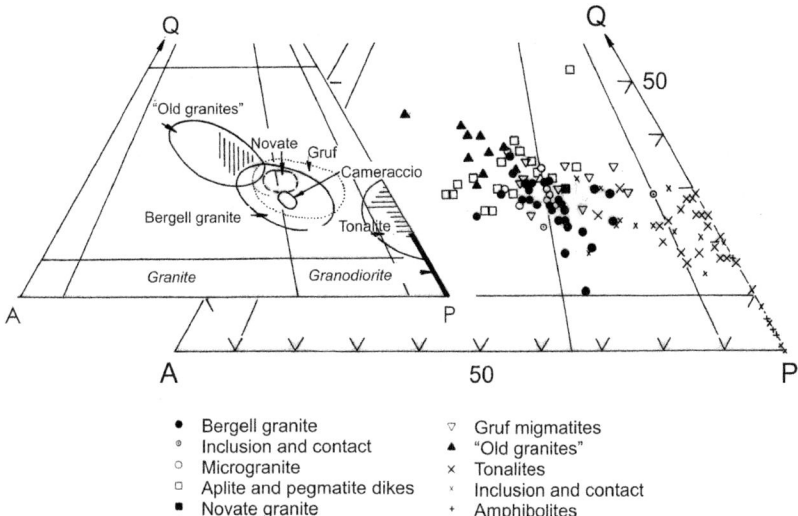

- Bergell granite
- Inclusion and contact
- Microgranite
- Aplite and pegmatite dikes
- Novate granite
- Gruf migmatites
- "Old granites"
- Tonalites
- Inclusion and contact
- Amphibolites

Fig. 3.7 Ternary Alkali feldspar-Plagioclase-Quartz (APQ) diagram as representation of composition of granitic rocks in the Bergell Alps (modified from Wenk et al. 1977)

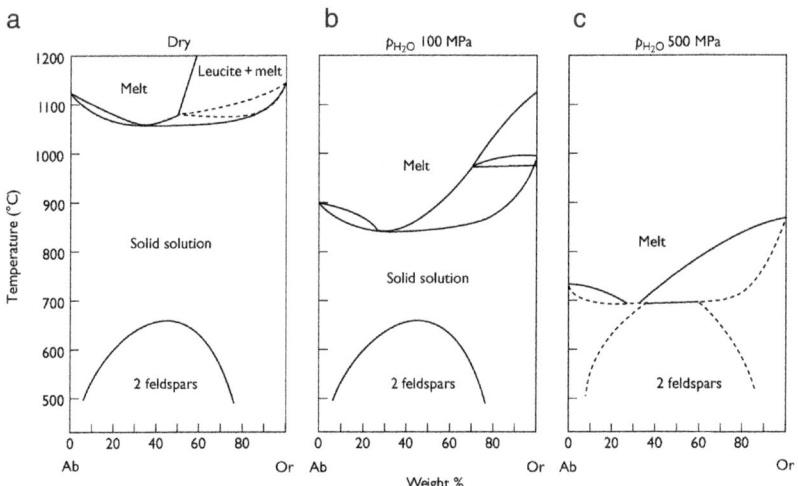

Fig. 3.8 Water plays a critical role for melting granitic rocks. Albite (Ab)-orthoclase (Or)-temperature phase diagrams with different water contents (after Bowen and Tuttle 1950)

Fig. 3.9 a View from Tombal above Soglio into Val Bondasca with Sciora group (center), Pizzo Cengalo and Pizzo Badile (on the right). What is white (covered by snow) is Bergell granite. The rest are rocks of Gruf and Tambo nappes. **b** View on Pizzo Cengalo (left) and Pizzo Badile (right). All is granite

3.3 Microstructure of Rocks, Their Composition and Formation Conditions

In the field, with a hammer and even a hand lens, it is often difficult to identify rocks and their mineral content as described in Sect. 3.1. Petrographers study the composition of rocks with an optical microscope, or, for more information about the chemical composition, with an electron microscope in laboratories. With these instruments, it is easy to recognize the minerals. Here we show a few pictures of thin sections observed with an optical microscope (Fig. 3.10) and discover some fascinating details of Bergell rocks which you would never expect on a hike in the field.

From a hand specimen we need to prepare a 30 µm thin section mounted on a glass slide which is then observed with a petrographic microscope in transmission, using polarized light. The granite in Fig. 3.6 looks quite different when observed with a microscope (Fig. 3.10a). Alkali feldspars (A) are finely twinned with a cross-hatched pattern, typical for triclinic microcline transforming from monoclinic orthoclase during cooling. Plagioclase (P) is zoned which shows that the composition (Albite $NaAlSi_3O$—Anorthite $CaAl_2Si_2O_8$) has changed during crystallization from the melt. The center is darker and richer in Ca (higher temperature) than the periphery (richer in Na at lower temperature). Quartz (Q) shows

3.3 Microstructure of Rocks, Their Composition and Formation Conditions

Fig. 3.10 Thin sections viewed with a petrographic microscope. **a–c** Use cross-polarized light, **d** plane polarized light. **a** Bergell granite from Val Forno. Alkali feldspar (A) shows fine cross-hatched twinning indicative of microcline. Plagioclase (P) displays zoning, dark Ca-rich in center and light Na rich around it. Quartz (Q) is quartz with undulatory extinction indicative of deformation. Bi is biotite. **b** Andalusite (An)-kyanite (Ky)-sillimanite (Si)-garnet (G) schist from Cataeggio (Val Masino) representing a triple point assemblage. **c** Amphibolite from Cavloccio with albite (Ab) and anorthite (An) coexisting. Bi is biotite. **d** Brown stilpnomelane crystals in muscovite-quartz schist from Cavril/Maloja. Scale in **d** applies to all micrographs

an undulatory structure typical for microscopic deformation. Biotite (Bi) is easily recognizable (brown), but there is also muscovite (yellow/green at the bottom).

The association of various minerals is linked to overall chemical composition, temperature and pressure conditions. Good examples from the Bergell are aluminosilicates crystallizing in schists dominantly composed of plagioclase, biotite and muscovite but with accessories. There are three minerals andalusite, sillimanite and kyanite (Table 3.2) that have the same chemical composition Al_2SiO_5 but a different crystal structure ("polymorphism") and therefore look different.

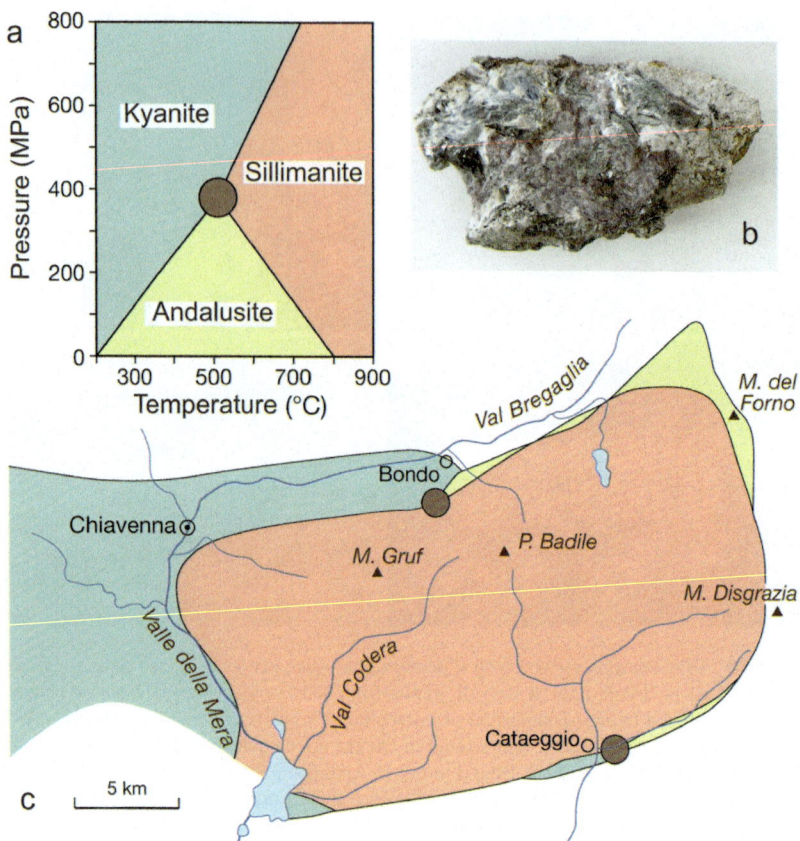

Fig. 3.11 **a** PT phase diagram Al_2SiO_5 with andalusite-kyanite and sillimanite. **b** Sample of andalusite (pink) and kyanite (blue) from Ciresc/Bondo. **c** Distribution of the three minerals around the Bergell granite. Triple points with all polymorphs coexisting are dark circles (after Wenk et al. 1974)

Figure 3.11b shows a hand specimen with blue kyanite and pink andalusite collected at Ciresc above Bondo. Figure 3.11a is a pressure–temperature (PT) phase diagram of aluminosilicates which displays conditions at which these three minerals are thermodynamically stable. At low pressure andalusite crystallizes (yellow in Fig. 3.11a) and in fact andalusite is found extensively on the east side and the

3.3 Microstructure of Rocks, Their Composition and Formation Conditions

roof of the Bergell massif around Monte del Forno (Fig. 3.11c). At higher pressure, kyanite is stable (green in Fig. 3.11a). Kyanite can be found in the deeper nappes in the west, as far as Ticino (Fig. 3.11c). At high temperatures (orange in Fig. 3.11a), sillimanite crystallizes in schists from the lower contact of the granite and Gruf migmatites (Fig. 3.11c).

PT conditions can be estimated from the occurrence of these minerals. In the Bergell region there are two extraordinary points where all three polymorphs occur together and therefore PT conditions are defined (T = 520 °C, P = 400 MPa). This point is called a triple point (Fig. 3.11a) where the three minerals are in thermodynamic equilibrium. A co-existence of these three minerals has not been documented anywhere else in the world except in the Bergell Alps with two triple points, one south of Bondo, near Ciresc, and a second one east of Cataeggio in Val Màsino (Fig. 3.11c). Figure 3.10b is a thin section image of the three minerals that crystallize in equilibrium under these conditions at the millimeter scale in a schist near Cataeggio (Wenk et al. 1974).

Another unique observation relates to plagioclase feldspars albite ($NaAlSi_3O_8$) and anorthite ($CaAl_2Si_2O_8$) in amphibolite from the contact zone with the Bergell granite at Cavloccio (excursions 4.1.1 and 4.2.1). Amphibolite is a Fe–Mg–Ca rich rock, mainly composed of hornblende and plagioclase. The temperature–composition phase diagram of plagioclase (albite-anorthite) is complicated (Fig. 3.12). At high temperature, but below the melting point, there is a field of continuous compositions between albite (Ab) and anorthite (An), called a solid solution. At intermediate temperatures, during cooling, plagioclase segregates into different compositions. Depending on composition pairs of albite-andesine (Perthite Pe), andesine-bytownite (Bøggild Bø) and bytownite-anorthite (Huttenlocher Hu) are present, often with lamellar structures as in the classical labradorite from Labrador/Canada with iridescence. But at lower temperatures (below 400 °C), albite and anorthite are in equilibrium (Fig. 3.12) and this corresponds to the conditions at Cavloccio at the NE contact of the granite when amphibolite recrystallized. In Fig. 3.10c one can recognize crystals of albite and anorthite co-existing, with no indication of exsolution (Wenk 1979). Such observations also require an optical microscope. Remember this unique mineral assembly when you look at the amphibolite on your fieldtrip (excursions 4.1.1 and 4.2.1).

The three microscope images (Fig. 3.10a–c) have been recorded with cross-polarized light. Figure 3.10d used plane polarized light and we are looking at a greenschist facies schist from Cavril/Maloja. The brown needle-like mineral is stilpnomelane ($KFe_8(Si_{11}AlO_{28})(OH)_8 \cdot 2H_2O$) intergrown with muscovite and quartz. Also this mineral would be impossible to identify in a hand specimen.

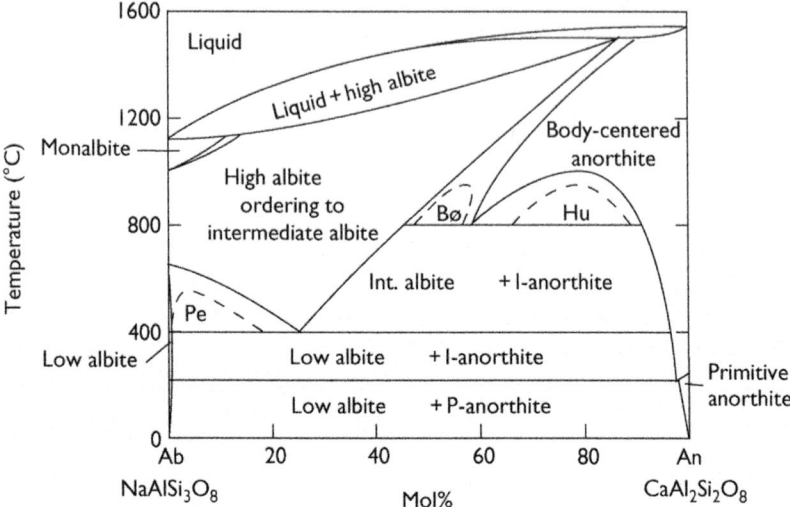

Fig. 3.12 Temperature–composition phase diagram of plagioclase: albite (Ab)-anorthite (An). At low temperature albite and anorthite are in equilibrium (after Smith and Brown 1988)

These examples introduce you to some of the complexities of geological studies and the rewards that you harvest. On your fieldtrips keep in mind that you are walking through a complex and fascinating space of compositions, temperature and pressure conditions that changed during the geological history.

3.4 The Recent History: Glaciers and Moraines

As mentioned earlier the only sediments in Val Bregaglia are quaternary deposits. In terms of area, they are quite significant, about 10% alluvium deposited by rivers, 25% talus, rock falls and landslides on mountain slopes, 25% moraines, compared to 40% rock outcrops. On the Lunghin Pass is an important triple point, where the catchment areas of the Rhine (North Sea), the Inn (Danube, Black Sea) and the Maira (Po, Adriatic Sea) meet (excursion 4.2.2).

However, this morphological triple point was not always at the same place as Fig. 3.13a shows with a reconstruction of the glaciation during the Last Glacial Maximum (LGM) (e.g. Bini et al. 2009). The Last Great Ice age period (Riss/

3.4 The Recent History: Glaciers and Moraines

Würm) lasted from 115,000 to 10,000 years before present (bp). The LGM in the Alps took place ~25,000 years ago which has been investigated in detail (e.g. Jäckli 1962; Hantke 1983), particularly recently with isotope methods, investigating boulders and variations in ^{10}Be and ^{36}Cl contents (e.g. Ivy-Ochs and Kober 2008). Contrary to ^{238}U ^{206}Pb isotopes which date millions and billions of years, ^{10}Be ^{36}Cl methods investigate variations of thousands of years. Applying them, the extent of glaciers on the S-side of the Alps can be mapped in great detail (e.g. Gianotti et al. 2015; Kamleitner et al. 2022).

Glaciers from Piz Bernina in the Engadine moved partly to the west and inside moraines on the Val Bregaglia north side you can find numerous boulders of Err-Julier granite. In the Engadine, Plan Camfer (S of Bivio), Jufer Alp and Val Bregaglia, the glaciers of the Last Ice Age were essential for creating the present morphology.

Since the Würm Ice Age, there have been large temperature fluctuations that can best be quantified by studies of the ice compositions in Greenland (e.g. Alley 2000; Fig. 3.13b). There was a cold period around 700 bp-before present ("Little Ice Age"). Around 1000 bp there was a medieval warm period with temperatures similar as today, and around 2000 bp a Roman climate optimum took place when Hannibal's elephants crossed the Alps and Bergell glaciers may not have existed.

After glaciers expanded during the "Little Ice Age" they began retreating again during the last 100 years. As an old photograph documents, in 1935 the Forno Glacier extended more than 2 km further north (Fig. 3.13c) compared with the current position (2022, Fig. 3.13d). Impressive is the Dufour topographic map from 1857 where the Forno Glacier reached almost as far as Plan Canin (Fig. 3.13e). Glacial retreat in the last 120 years has averaged about 30 m per year (e.g. Maurizio 1982). In the last 50 years, anthropogenic global warming has certainly contributed to this, but as Fig. 3.13b documents the Earth was subjected to large temperature changes long before.

Spectacular glacial mills of the Bergell Glacier can be found west of Maloja (Fig. 3.14a, 773.0/141.8, excursion 4.1.2) and east of Chiavenna (see Fig. 4.7f, excursion 4.1.6). They form on bedrock when river water twists pebbles around and grinds holes. Glaciers also create polished flat surfaces by pushing rock boulders over outcrops. Good examples can be observed at Orden (Fig. 3.14b; excursion 4.1.1).

At the Chiavenna outcrop there are also large erratic blocks of Bergell granite on these surfaces, far away from their original location (Excursion 4.1.6, see Fig. 4.7g). This glacial polish often has lineations. If you strike with your hand back and forth, one direction is smooth and the other direction is rough and you can determine the direction in which the glacier moved.

Fig. 3.13 **a** View on SE Alps during the Last Glacial Maximum (Bini et al. 2009). **b** Temperature variations since the last Ice age (after Alley 2000). **c–e** Forno glacier. **c** 1935, **d** 2022, **e** Dufour topographic map, 1857. Look at the extent of glaciers, especially Forno which reached almost to Plan Canin. Also interesting are names of mountains: Cma. di Tschingel (now Pizzo Badile), P. Trubinasca (now Pizzo Cengalo), M. d'Oro (now Monte Forno)

3.4 The Recent History: Glaciers and Moraines

Fig. 3.14 a Glacial mills at Maloja. **b** Large granite blocks of Bergell granite across the main valley road (Sasc Taca below Stampa). **c** Glacial polish with striations at Maloja/Orden. **d** Miocene Gonfolite conglomerate near Chiasso with boulders of Bergell granite and tonalite

On the north side of Val Bregaglia moraines contain Julier granite (Durbegia, Tombal, Soglio), Bergell tonalite (Durbegia) and rarely porphyroclastic Bergell granite (Dasciun, Durbegia). There are numerous smaller secondary local moraines.

Large postglacial alluvium can be found in the valley floor of Val Bregaglia, especially below Casaccia where gravel is currently being excavated, mainly for the production of concrete. Compared with the flat morphology in the Upper Engadine, east of Maloja, there is a steep elevation drop of 300 m towards the west, probably caused by the erosion-resistant Margna gneiss. On both sides of Val Bregaglia there are many talus slopes and local landslides.

Interesting features are huge blocks of Bergell granite, visible for example on the main valley highway at Sasc Taca below Stampa (Fig. 3.14c) and at Creista (Borgonovo). They are barely rounded and were probably deposited in a late phase of glaciation from the southside, where Bergell granite is exposed at high elevations, thrust over gneisses of Gruf migmatites and the Tambo nappe.

It should be mentioned here that erosion in the Bergell Alps is not mainly due to glaciers. Long before the Ice Age there was rapid erosion and pebbles of first tonalite (~ 27 My) and later Bergell granodiorite (~ 20 My) were deposited in late Oligocene early Miocene. Gonfolite conglomerates in the South-Alpine foreland near Como and Chiasso formed before Novate granite intruded (~ 20–25 My) (Fig. 3.14d; e.g. Di Capua et al. 2015).

3.5 Landslides

Everywhere in the Bergell and side valleys there is evidence of large landslides. We already briefly mentioned the most significant landslide in the region that on September 4, 1618 destroyed the large and prosperous town of Plurs (Piuro) which at that time belonged to Graubünden (Fig. 3.15a). On the south side of the valley above Plurs there were hundreds of mines and quarries excavating talc-olivine schists used for the production of valued stone pottery (Sect. 3.6.1)—the source of prosperity in Plurs. Mining in these ultramafic rocks may have been the cause of the landslide. It could also have been natural causes because these rocks are unstable as can be witnessed by the rockfalls such as Ganda Rossa above Ciresc/Bondo or Foppate above Villa di Chiavenna. On the small road from Villa di Chiavenna to Savogno (N-side of the valley) there are impressive outcrops of gneiss with brittle fractures and displacements (Fig. 3.15b) displaying rocks that are not very stable. The geology of this area was recently investigated in detail (Pigazzi et al. 2022).

Where the old town of Plurs with magnificent palaces once stood are now meadows and farms (Fig. 3.15c). Only one palace, Palazzo Vertemate-Franchi, on the northside of the town, survived (excursion 4.1.6). The death toll in the Plurs landslide in 1618 was over 1000.

Plurs it not the only landslide. A rockfall occurred under Piz Lunghin in 1970 and filled a large part of the valley directly above Casaccia (Fig. 3.16a). A similar much larger landslide happened in 1673, covering the village and surrounded some houses with 2 m thick debris so that what was once the ground floor now became a basement (Fig. 3.16b). Geologically there are many old landslides such as the one in Val Maroz (excursions 4.1.3 and 4.2.4; Fig. 3.16c) or a large slide

3.5 Landslides

Fig. 3.15 **a** Piuro before and after 1618 landslide (Scheuchzer 1723, see also display in Palazzo Vertemate-Franchi, Prosto). **b** Brittle faulting with displacements in rocks of the Tambo nappe along the road to Savogno. **c** View of Piuro landslide in 2021

over Muntac. In the summer of 2015, for over a month, there were rock falls occurring approximately every half hour under Piz dal Märc above Soglio, each time raising dust clouds (Fig. 3.16d).

In 1988 there was a landslide on the south side of Val Bregaglia, in Val Torta, that almost reached Vicosoprano (Fig. 3.16e) and in 2005 yet another in Val Scalota where the valley road was interrupted for a few days (Fig. 3.16f). These slides brought Bergell granite onto the valley floor.

Fig. 3.16 Landslides in Val Bregaglia. **a** Casaccia 1970 landslide from Piz Lunghin. **b** House in Casaccia partially covered by 1673 landslide. **c** Large landslide in Maroz-Denc. **d** Rockfall with bright dust above Soglio, 2015. **e**–**f** Landslides on the southern side of the Valley: **e** near Vicosoprano Val Torta, 1988 and **f** Val Scalota, 2005

3.5 Landslides

A small rockfall in 2011 from Pizzo Cengalo in Val Bondasca massively increased in summer 2017, melting parts of the Bondasca glacier which lay underneath moraines. This caused a huge landslide covering the upper Val Bondasca on August 23, 2017 (Fig. 3.17a, b). Debris came all the way to the town of Bondo but spared the old village and Palazzo Salis as they had been carefully built far away from the river (Fig. 3.17c). Only newer structures, closer to the river were destroyed (Fig. 3.17c, d).

There were also large landslides in neighboring Val Spluga where on both sides of the valley villages were destroyed, especially in Cimaganda (1533/34).

Fig. 3.17 Pictures of the Cengalo landslide, 2017. **a, b** Val Bondasca before and after the landslide (Google Earth 2015 and 2022), **c, d** Bondo (courtesy Comune di Bregaglia)

In 2012 the road was again interrupted by a rockfall for a month and in 2018 there was yet another big rock fall just above the Santuario di Gallivaggio. The church was miraculously spared (excursion 4.1.8).

The main causes of landslides are related to geomorphology, with a steep 2000 m difference in altitude between the valley floor and summits. This creates gravitational stresses. In part, the climate is responsible. Many landslides are associated with heavy thunderstorms. Added to it are the complex heterogeneous rock formations.

3.6 Mining, Quarrying and Water

In spite of the great variety of rocks, at least the upper Bergell was never a large mining area. In ancient times Romans mined gold in Val Aurosina ("aurum" Latin for gold) south of Piuro/Santa Croce. Historically, the most important product was "soap stone" or *lavez*, as we will mainly refer to in this book. It is an ultramafic talc-olivine schist that was mined for centuries around Piuro (Plurs). Locally, marble was used to produce lime for cement. Roof tiles were made from gneiss and quarries at Promontogno and below Soglio still produce gneiss slabs. Today, gravel pits at Casaccia are important for building materials. There is some evidence for copper extraction from local ore deposits. Water is a resource for hydroelectric energy with power plants at Löbbia and Casaccia.

3.6.1 Lavez (Olivine-Talc Schist)

Granite and gneiss have their origin in the continental crust. But rocks such as peridotite, serpentinite, talc-olivine schist came originally from the Earth's mantle, deep below the continental crust (500 km) and relatively shallow below the oceanic crust (50 km). The rocks are mafic to ultramafic (little SiO_2, no quartz) but rich in Mg and Fe. Parts of the mantle are mixed with the crust during tectonic processes. Where water is present, olivine transforms into serpentine or talc.

An ultramafic olivine-talc zone extends from Chiavenna in the west to Val Bondasca in the east (e.g. Schmutz 1976). This rock, also known as lavez, laveggio, Topfstein or pietra ollare (soapstone) has unique properties. Here we will use the name "lavez". It is compact and because of the soft talc it is easily workable and can be carved into bowls and cooking pots. There are already prehistoric lavez relics (Museum Sondrio) and lavez was intensively mined by Romans in

3.6 Mining, Quarrying and Water

the area of Chiavenna-Piuro and Valmalenco for the production of lavez pots, statues and troughs, leading to great prosperity in the region, such as the town of Plurs with its magnificent palaces (Fig. 3.15a, top).

The lavez preparation was described by Scheuchzer in 1723 (Fig. 3.18a). Due to soft talc it can be carved with steel blades. A lavez workshop from the last century in Valbrutta/Malenco is exhibited in the Ciäsa Granda Museum in Stampa (Fig. 3.18b). The largest documented lavez mine in Val Bregaglia lies above Prosto and is 240 m deep (752.6/131.9). You enter through a small fracture in the rock outcrop (Fig. 3.18c). Lavez of high quality is usually limited to narrow zones which are then exploited. Round blocks are carved out (Fig. 3.18d), then processed in workshops. From large blocks of talc-olivine schist several pots can be extracted, one inside the other, leaving a small block in the center that has been used as a cobble stone for roads (Fig. 3.18e). Today there is still the workshop of Roberto Luchinetti in Prosto and should be visited on excursion 4.1.6. Currently the only operating quarry for lavez material is in Piuro (754.0/131.9, Fig. 3.18f), but between Chiavenna and Piuro there were hundreds of historic lavez quarries and mines.

There was also a small quarry in Predacia/Bondasca near Gerp with a year 1773 (763.9/132.35, Maurizio 1972) and near it, in Ciresc above Bondo, you find round holes in lavez rocks, perhaps prehistoric and used for grinding chestnuts (excursion 4.2.8, see Fig. 4.20g).

Lavez mining also took place in the Ticino (e.g. Girlanda and Pfeifer 2020) and Valtellina/Valmalenco (e.g. Lurati 1970), often in hard-to-reach places in the mountains. In the Chiavenna zone there is also a fibrous variety of serpentine near Uschione, so-called chrysotile asbestos (Fig. 3.2g). It has long been applied as a heat-resistant insulation material. Today, chrysotile is no longer used because the fibers, when inhaled, cause lung cancer.

3.6.2 Lime Production

Historically important was the production of lime for construction of buildings. Lime (CaO) is produced by heating limestone or marble to convert it into lime at ~1000C: $CaCO_3 \rightarrow CaO + CO_2$. A reaction with water produces calcium hydroxide, $CaO + H_2O \rightarrow Ca(OH)_2$ which can be used like cement for construction. Over long periods of time, calcium hydroxide reacts with air and becomes calcite again.

Fig. 3.18 **a** Lavez manufacturing (Scheuchzer 1723). **b** Lavez workshop from Valbrutta, Malenco. Museum Stampa (Loan of Rhätisches Museum, Chur) (courtesy W. Hunkeler). **c** Small entrance to the largest lavez mine (240 m depth) above Prosto. It dates back more than 1000 years. **d** Zone in this quarry where large blocks were obtained for pot production. **e** Set of pots from a single block with remaining round insert often used as cobble stones. **f** Only lavez quarry in operation in 2021, behind Piuro

3.6 Mining, Quarrying and Water

Old lime kilns can still be found in many places such as northeast of Isola (777.9/143.3), Crap da Chürn-Plaun da Lej on Lake Sils (Fig. 3.19a, 775.9/143.4), Blaunca/Grevasalvas (774.3/143.4), Casaccia (Su l'Aua, 771.0/140.0), Vicosoprano-Crot (769.2/136.3), Stampa-Mulin (765.2/134.4, with quarry in an 1892 picture by Augusto Giacometti Fig. 3.19b) which is difficult to find today, southwest of Tombal (Fig. 3.19c, 762.0/135.1) and more remote places including Plän Vest (762.1/135.4), Castel (765.2/136.8), Larec (766.4/136.3), Zocheta (Fig. 3.19d, 766.0/135.2) and Cavi (760.1/135.3). It is worth exploring these ancient lime kilns in the field.

Fig. 3.19 Lime kilns at **a** Plaun da Lej (Crap da Chürn). **b** Painting of Augusto Giacometti (1892) with lime production in Stampa, above brown house (courtesy Centro Giacometti). **c** Furnace at Tombal (Canun dal Trocc). **d** Furnace on the meadow Zocheta above Durbegia

3.6.3 Quarries of Gneiss and Gravel Pits

Granite recrystallizes under tectonic stress and strain and becomes gneiss. The Tambo gneiss extends from Stampa-Promontogno-Soglio towards the northwest to Piz Tambo near the Splügen Pass. Above Chiavenna, in the river Liro, the granitic texture is preserved (excursion 4.1.8, see Fig. 4.8g). In the Bergell, this old Tambo granite was severely deformed and transformed to gneiss. Mica reorients itself and this forms the cleavage. A variety with an extraordinary cleavage is mined in quarries below Soglio (Fig. 3.20a) and near Promontogno (Fig. 3.20b). Perfectly flat plates up to 2 m in length occur (Fig. 3.20a) and are used for the production of stone tables or roof plates. Large quarries existed since the eighteenth century, especially in Promontogno near La Porta, below Nossa Donna and below Soglio, as described in old archives of the municipalities of Bondo and Soglio (Maurizio 1972). The grey stone roofs of houses such as those in Soglio (Fig. 3.20c) are made of this Tambo gneiss.

There is a large gravel pit at Casaccia (Kieswerk Casaccia AG) (Fig. 3.20d) in the alluvial plane. Gravel and sand are used for the production of concrete and road maintenance. In the gravel plant, old demolition material is also made usable again.

3.6.4 Metal Ores

Although we mentioned that Romans briefly explored for gold in Valle Aurosina, there is at least currently no good evidence for extensive historic gold exploration. Interestingly, on the old Dufour map sheet 10 (1857) we find the name M. d'Oro for what is currently called Monte del Forno (Fig. 3.13e), and Monte d'Oro is now the name of a mountain east of Muretto Pass in Italy. Maurizio (1972) describes that this is the legend of a farmer from Valmalenco who spread a story about gold at the beginning of the nineteenth century, but it could never be confirmed.

There are some manganese and copper deposits south of Monte del Forno that were briefly mined and processed in furnaces at Plan Canin—hence the name "Forno" (e.g. Ferrario and Montrasio 1976; Peretti and Köppel 1986).

Another old mine with Cu and Fe ores is at the foot of Mott Scalotta/Tavretga southwest of Bivio. These metal deposits probably correspond to "black smokers" of oceanic crust with chalcopyrite $CuFeS_2$ (Fig. 3.21a) and malachite $Cu_2(CO_3)(OH)_2$ (Fig. 3.21b).

3.6 Mining, Quarrying and Water

Fig. 3.20 **a** Dolfo Schumacher in Soglio quarry with large plates of Tambo gneiss. **b** Promontogno gneiss quarry, 2021. **c** Stone roofs in Soglio. **d** Gravel pits in Casaccia. Piz Lunghin in background

Above Casaccia on a flat spot in the forest called Mota Farun slags with copper traces have been found (Fig. 3.21c). With an electron microscope you can identify a range of "minerals" produced in the kiln: native copper, dendrites of pentlandite $FeNiS_2$, bornite Cu_5FeS_4 and pyrrhotite FeS (Fig. 3.21d). The composition and morphology of these slags is analogous to slags from the Oberhalbstein attributed to the late Bronze Age (Reitmaier-Naef et al. 2015). The minerals from which copper was extracted are chalcopyrite and malachite.

Fig. 3.21 Old copper mine Alp Tgavretga/Septimer Pass with **a** chalcopyrite and **b** malachite. Old copper production Motta Farun above Casaccia with **c** slags and **d** a scanning electron microscope image with native copper (Cu), and dendritic pentlandite (Pt) in a matrix of bornite (Bo) and pyrrhotite (Py) (after Wenk et al. 2020)

3.6.5 Water/Electricity

Water is an important component of the Earth. Most sedimentary rocks were formed in water, for example limestones, sandstones and shales. In the Bergell region they have been transformed into marble, quartzite and schists during metamorphism. Some of the water was preserved in metamorphic minerals, e.g. muscovite $KAl_2[Si_3Al_2O_{10}](OH)_2$, serpentine $Mg_3[Si_2O_5](OH)_4$, tremolite $Ca_2Mg_5[Si_8O_{22}](OH)_2$, epidote $Ca_2Fe^{3+}Al_2SiO_4(Si_2O_7)O(OH)_2$ (Table 3.1). Water is also a crucial component of the main synthetic mineral in cement, tobermorite $Ca_5Si_6O_{16}(OH)_2 \cdot 4H_2O$, known in materials science as C-S-H.

3.6 Mining, Quarrying and Water

Rain, snow and glaciers are important for geological processes on the surface of the Earth, such as landslides, moraines and erosion. Without water the Bergell would be entirely different. Currently the average precipitation in the Bergell is about 80 cm/year and it has changed little over 150 years (Fig. 3.22a).

Today, water is also by far the most important natural raw material of the Bergell. In 1955–1959, the Electricity Works of the City of Zurich (EWZ) built a large dam in Val Albigna with a reservoir volume of 79×10^6 m^3 (Fig. 3.22b) and numerous tunnels for water transport such as Albigna-Murtaira (5 km), Plan Canin-Murtaira (3 km), Murtaira-Löbbia (1.5 km), Löbbia-Castasegna (12 km) and Bondasca-Bondo (3 km) were constructed. The total catchment area is about 115 km^2, which corresponds to an annual water volume of $\sim 10^8$ m^3 or 10^{11} L.

Albigna and Forno/Plan Canin are at ~ 2000 m. From this elevation water flows in a tunnel to the electric power facility at Löbbia (1400 m, Fig. 3.22c) where electricity is produced and then continues in another tunnel to the facility at Castasegna (700 m). About half of the precipitation in Val Bregaglia is used and produces on average 440 GW hours per year which is about 20% of the electricity consumption in the city of Zurich. With currently ~ 30 employees, EWZ is the most important industry in the valley. EWZ also operates a cable car from Pranzaira to the Albigna dam which we will use on excursions 4.1.6 and 4.2.7.

The tunnels were also very important for geology. Two tunnels Plan Canin-Murtaira and Albigna Murtaira pass through the contact zone of the Bergell granite and have been studied in detail (Weibel and Locher 1964). Particularly interesting is the tunnel from Löbbia to Castasegna where the surface is largely covered by quaternary deposits and the tunnel provides valuable information about the underlying rocks of the Tambo, Suretta, Adula and Lizun nappes (Fig. 3.23).

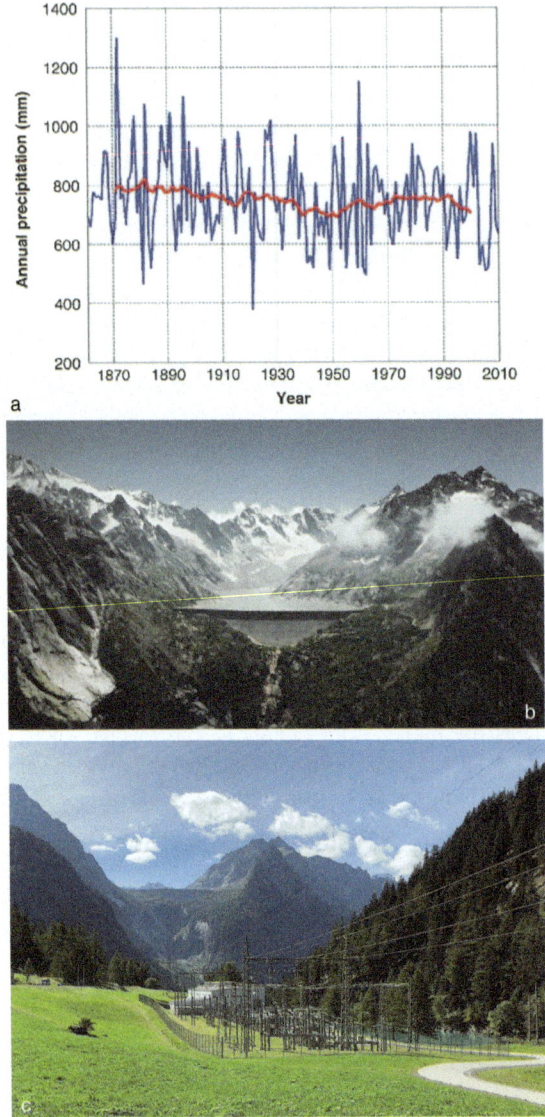

Fig. 3.22 a Changes in annual precipitation measured in Samedan (1861–2010) (after Zekollari et al. 2014). **b** EWZ Albigna dam built in 1957. **c** Löbbia hydroelectric power station and behind the Albigna dam (**b** courtesy of EWZ)

3.6 Mining, Quarrying and Water

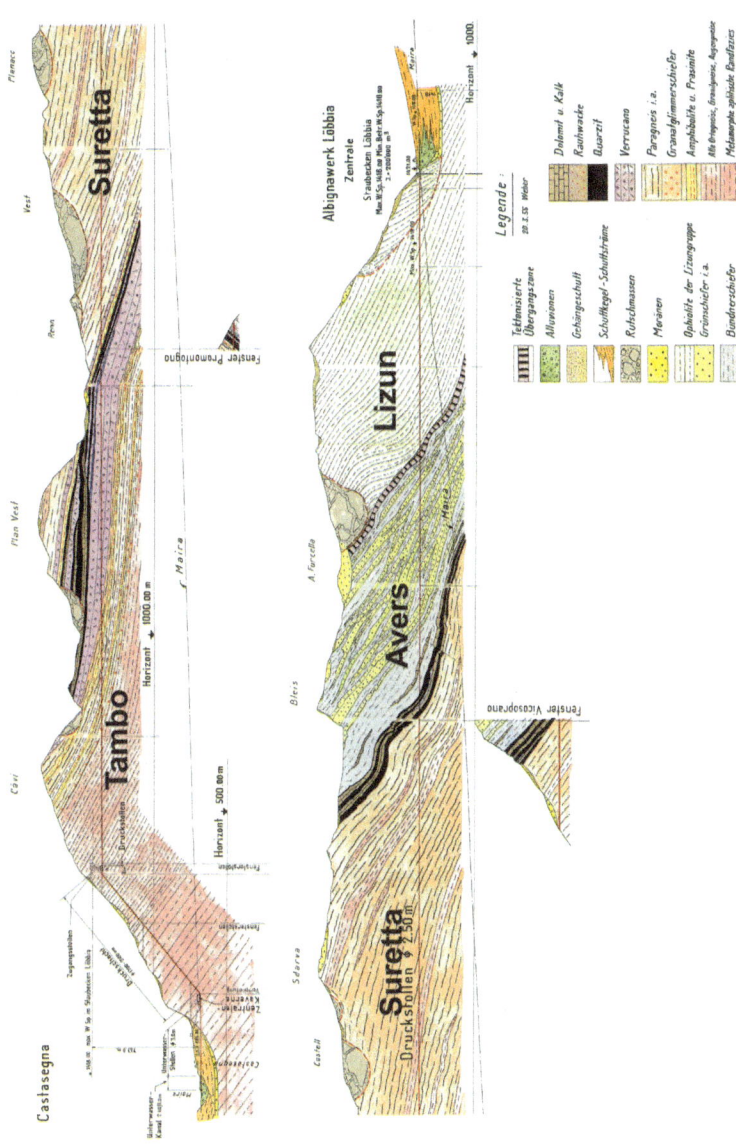

Fig. 3.23 Geological profile Löbbia-Castasegna with a stack of Pennine nappes along the EWZ tunnel. Important nappes are highlighted. Black is Triassic quartzite (courtesy of EWZ)

References

Alley RB (2000) Ice-core evidence of abrupt climate changes. PNAS 97(4):1331–1334

Bedogné F, Maurizio R, Montrasio A, Sciesa E (1995) I Minerali della Provincia di Sondrio e della Bregaglia Grigionese. Bettini, Sondrio, 275 pp

Bini A, Buoncristiani J-F, Couterrand S, Ellwanger D, Felber M, Florineth D, Graf HR, Keller O, Kelly M, Schlüchter C, Schoeneich P (2009) Die Schweiz während des Letzteiszeitlichen Maximums (LGM). Bundesamt für Landestopografie Swisstopo, Wabern

Bondi M, Griffin WL, Mattioli V, Mottana A (1983) Chiavennite, $CaMnBe_2Si_5O_{13}(OH)_2$ $2H_2O$, a new mineral from Chiavenna (Italy). Am Mineral 68:623–627

Bowen NL, Tuttle OF (1950) The system $NaAlSi_3O_8$-$KAlSi_3O_8$-H_2O. J Geol 58:498–511

Di Capua A, Vezzoli G, Cavallo A, Gropelli G (2015) Clastic sedimentation in the Late Oligocene Southalpine Foredeep: from tectonically controlled melting to tectonically driven erosion. Geol J 51(3):338–353

Ferrario A, Montrasio A (1976) Manganese ore deposit of Monte del Forno. Its stratigraphic and structural implications. Schweiz Mineral Petrogr Mitt 56:377–385

Garlick S (2014) Pocket guide to rocks and minerals of North America. National Geographic, 184 pp

Gianotti F, Forno MG, Ivy-Ochs S, Monegato G, Pini R, Ravazzi C (2015) Stratigraphy of the Ivrea morainic Amphitheatre (NW Italy): an updated synthesis. Alpine Mediterr Quat 28:29–58

Girlanda F, Pfeifer H-R (2020) La pietra ollare nelle Centovalli e Terre di Pedemonte. Una ricerca tra storia, etnografia e scienza. Terretre 75:5–9

Grünenfelder M, Stern TW (1960) Das Zirkon-Alter des Bergeller Granits. Schweiz Mineral Petrogr Mitt 40:253–259

Hantke R (1983) Eiszeitalter: Die jüngste Erdgeschichte der Schweiz und ihrer Nachbargebiete. Westliche Ostalpen mit ihrem bayerischen Vorland bis zum Inn-Durchbruch. Volume 3. Ott Verlag, 730 pp

Ivy-Ochs S, Kober F (2008) Surface exposure dating with cosmogenic nuclides. Quat Sci J 57:179–209

Jäckli H (1962) Die Vergletscherung der Schweiz im Würmmaximum. Eclogae Geol Helv 55:285–294

Kamleitner S, Ivy-Ochs S, Monegato G, Gianotti F, Akcar N, Vockenhuber C, Christl M, Synal H-A (2022) The Ticino-Toce glacier system (Swiss-Italian Alps) in the framework of the Alpine Last Glacial Maximum. Quat Sci Rev 279:107400

Lurati O (1970) L'ultimo laveggiaio di Val Malenco. In: Sterbendes Altes Handwerk. Schweizerische Gesellschaft für Volkskunde, Basel, p 24

Maurizio R (1972) Indagini su vecchie cave e miniere in Bregaglia. Quad Grigioni Ital 41:1–71

Maurizio R (1982) In forte regresso il ghiacciaio del Forno. Almanacco del Grigioni Italiano, pp 98–116

Maurizio R, Meisser N (1993) Neue Mineralien des Bergells (Schweiz-Italien). Schweiz Strahler 9(11):525–557

Maurizio R, Weibel M (1982) Die Mineralien des Bergells. Mineralienfreund 20(4):81–100

Meyer J (2017) Gesteine einfach Bestimmen: Der Bestimmungsschlüssel. Haupt Verlag, 140 pp

Peretti A, Köppel V (1986) Geochemical and lead isotope evidence for a mid-ocean ridge type mineralization within a polymetamorphic ophiolite complex (Monte del Forno, North Italy/Switzerland). Earth Planet Sci Lett 80(3–4):252–264

Pigazzi E, Bersezio R, Morcioni A, Tantardini D, Apuani T (2022) Geology of the area of the Piuro 1618 event (Val Bregaglia, Italian Central Alps): the setting of a catastrophic historical landslide. J Maps. https://doi.org/10.1080/17445647.2022.2057878

Reitmaier-Naef L, Rouven T, Casa PD (2015) Praehistorische Kupfergewinnung im Oberhalbstein. Miner Helv 36:35–54

Scheuchzer JJ (1723) Helveticus sive Itinera per Helvetiae Alpinas Regiones. In: Lugduni Batavorum MDCCXXIII

Schmutz H-U (1976) Der Mafitit-Ultramafitit-Komplex zwischen Chiavenna und Val Bondasca (Provinz Sondrio, Italien; Kt. Graubünden, Schweiz). Beitr Geol Karte Schweiz, Neue Folge 149, 73 pp

Smith JV, Brown WL (1988) Feldspar Mineralogy. Springer Verlag, Berlin, 828 pp

Weibel M, Locher T (1964) Die Kontaktgesteine im Albigna- und Fornostollen (nördliches Bergeller Massiv). Schweiz Mineral Petrogr Mitt 44:157–185

Weibel M, Gräser S, Oberholzer WR, Stalder HA, Gabriel W (1990) Die Mineralien der Schweiz, 5th edn. Birkhäuser-Springer, 224 pp

Wenk H-R (1979) An albite-anorthite assemblage in low-grade amphibolite facies rocks. Am Mineral 64:1294–1299

Wenk H-R (1983) Mullite-sillimanite intergrowth from pelitic inclusions in Bergell tonalite. N Jb Mineral Mh 146:1–14

Wenk H-R, Maurizio R (1978) Kutnahorite, a rare Mn mineral from Piz Cam (Bergell Alps). Schweiz Mineral Petrogr Mitt 58:97–100

Wenk H-R, Wenk E, Wallace J (1974) Metamorphic mineral assemblages in pelitic rocks of the Bergell Alps. Schweiz Mineral Petrogr Mitt 54:507–554

Wenk H-R, Hsiao J, Flowers G, Weibel M, Ayranci A (1977) A geochemical survey of granitic rocks in the Bergell Alps. Schweiz Mineral Petrogr Mitt 57:233–265

Wenk H-R, Yu R, Tamura N, Bischoff D, Hunkeler W (2020) Slags as evidence for copper mining above Casaccia, Val Bregaglia (Central Alps). Minerals 9:292

Zekollari H, Fürst JJ, Huybrechts P (2014) Modelling the evolution of Vadret da Morteratsch, Switzerland since the Little Ice Age and into the future. J Glac 60:1155–1168

Geological Excursions in the Bergell Alps

4

An important part of this geological guide are excursions in the Bergell area, mostly on marked trails (Fig. 4.1a). Easy trails are marked red-white, more technical ones blue-white. On trails we sometimes walk on cobblestones such as the old road into Val Maroz (Fig. 4.1b) but we are mainly on gravel and soil, not rock outcrops. A couple of fieldtrips require scrambling off trail. Be sure you don't get lost. Climbing in the Bergell mountains is a big attraction, even for small children (Fig. 4.1c). On some excursions you may cross a glacier and an ice axe is recommended (Fig. 4.1d). It is assumed that you are familiar with some mountaineering skills because geological outcrops are often not directly on the trails.

In the Alps hiking is never without danger: you can miss the trail, a thunderstorm with lightning can be dangerous (especially with a geological hammer in your hand), a steep snowfield must be crossed carefully. We divide the excursions into two categories: simple walks for all, including families with children, 1–4 h, not very strenuous; and larger more challenging hikes to higher elevations and options to climb some summits such as Monte del Forno, Piz Lunghin, Piz Duan, Piz Cam, Piz Casnil, Piz Cantun and Piz Salacina. Even if these climbs are not technically difficult, a rope is recommended, especially with children, and you need to consult guides such as Nigg (2004) or Meier and Alig (2006).

Most excursions are on the Swiss Topographic Maps 1:25,000 Val Bregaglia (1276) and Sciora (1296), and 1:50,000 S. Bernardino (267), Julierpass (268), Roveredo (277) and M. Disgrazia (278). There are also excellent topographic maps with paths available on the internet such as http://map.geo.admin.ch/ or https://mapcarta.com/17668508.

Fig. 4.1 **a** Marker at intersection of trails at Salacina junction. **b** Hiking on trails, such as the Roman-Medieval road into Val Maroz. **c** Here we show a small cliff in Durbegia where even children enjoy some climbing. **d** Ugo Ruinelli, a local guide and enthusiastic about minerals with an ice axe, backpack and hammer near a crevasse on the Forno glacier (courtesy W. Hunkeler).

For geological maps use the link (https://map.geo.admin.ch/?topic=geol&lang=en) and navigate to the Bergell region. For published maps in Switzerland see Peters (2005), Peters et al. (2008), Spillmann and Trommsdorff (2005), Wenk and Cornelius (1977) and in Italy Montrasio and Sciesa (1988), and Montrasio et al. (2005).

4 Geological Excursions in the Bergell Alps 67

For each excursion we provide a map with trail, important points [#] projected on a geological background. These ten geological maps with a legend are collected at the end of the book. The position of the maps is displayed as rectangles in an overview. For coordinates of places we use the system of Swiss topographical maps S–N/W–E, in km. For example the Alp Maroz Dora in Val Maroz is at coordinates 140.6/769.5.

We recommend that you download the SwissTopo App for your phone before embarking on any of the described tours. It is available at the Apple and Google-Play App Stores, with more information at https://www.swisstopo.admin.ch/en/maps-data-online/maps-geodata-online/swisstopo-app.html. For higher resolution geological maps, with more details about geological units go to the SwissTopo website.

For accessing the digital map data from your phone in the field you may choose to have either the standard topographic map or the geological map layer as your base layer. The topographic map is the default; to select the geological map, select the "layers" icon in the lower right corner of the map, then scroll right to 'Pro' and you will find Geological Map as one of the options. Note, this functionality was only added in 2023, so if you installed SwissTopo previously, you need to visit the App Store and update your version of SwissTopo (for free!).

Waypoints and tracks for the tours described in this book are available separately at https://www.tinyurl.com/BergellGeologyTours. You need to download these files to your phone. There are two files for each tour, one with the track for the tour and a second with waypoints, the geologic points-of-interest. For some users, it will be easiest to access the website (a GoogleDrive folder) directly from your phone and others may prefer to access the files on a computer, then email them to your phone. For iPhones, you need to email the files as attachments, then access your email through the native Apple mail app, not other mail apps (i.e. Outlook). See www.swisstopo.admin.ch/en/maps-data-online/maps-geodata-online/swisstopo-app/faq.html for details on the current best scheme to integrate the files with SwissTopo; this varies by phone manufacturer, phone operating system, etc. For an iPhone you press down on the attachment and select share, then scroll through the app options at the bottom of the screen until you find SwissTopo. The waypoints and tracks are then automatically integrated into SwissTopo for future use.

What do you need for an excursion? For geological purposes a rock hammer is good to have with you. Those interested in minerals will also need a hand lens and a small chisel. Bring good hiking boots and, depending on which excursion you take, an ice axe and a rope. With binoculars you can see details of remote mountains such as Piz Bernina, Monte Disgrazia and even Monte Rosa far to the west ... and closer by you may observe some ibex (Fig. 2.10b), chamois or an eagle. Rain protection is essential because thunderstorms are difficult to predict. Start the hike early. Thunderstorms usually come in the late afternoon and evening. Shelter during lightning. Lightning, particularly on summits, can even melt rocks and produce glass, so-called fulgurites. Bring enough food and snacks. Except for huts operated by the Swiss and Italian Alpine Clubs which provide meals part of the year, there are no restaurants on most routes. There are frequently springs and fountains along the way, but you should always have some water with you.

Be aware of warnings from municipalities: Some roads and paths are closed due to rockfall hazards, for example currently Val Bondasca. Before you start a trip, explore weather conditions. Access to quarries such as Soglio, Promontogno and Novate Mezzola may require a permit which should be easy to obtain from quarry managers.

The nineteen excursions are a start and, having become familiar with the local geology, you may want to adventure into different locations.

Accommodations

Capanna del Forno SAC, June–October, tel. 0041 81 824 32 82, www.fornohuette.ch
Capanna Albigna SAC, June–September, tel. 0041 81 822 14 05, albigna@bluewin.ch, www.albigna.ch
Capanna Sciora SAC, tel. 0041 81 822 11 38, sciora.cap@bluewin.ch, will reopen in 2025
Capanna Sasc Furä SAC, June–September, tel. 0041 81 822 12 52, info@sascfura.ch, www.sascfura.ch
Rifugio Luigi Brasca CAI, tel. 339 71 766 20, rifugiobrasca@inwind.it
Rifugio Tartaglione Crispo, http://www.rifugiotartaglione.com/
Bivacco Pedroni-Del Prà.
Rifugio Del Grande Camerini CAI, https://www.CASSovico.it/rifugio/rifugio.it

Camping

Mulina, Vicosoprano: www.bregaglia.ch/en/camp-mulina
Plan Curtinac, Maloja: www.engadin.ch/en/camping/camping-plan-curtinac/
Aqua Fraggia, Piuro: www.campingacquafraggia.com
La Torre, Sorico: www.campinglatorresorico.com.

4.1 Easy Hikes

4.1.1 Alp Cavloc: Tectonic Nappes and Contact Rocks (Map 1, waypoints #a1–#a10) (2–5 h)

This hike to Alp Cavloc can also be combined with a larger excursion to Val Forno (4.2.1). A map with a geological background can be found in Map 1 with a legend at the end of the book. It is best to park in the large parking lot 773.3/140.6 [marked #a1 on Map 1] just South of Maloja and near the large Orlegna dam that was built in 1971 to prevent flooding during large precipitations as in 1927, which destroyed part of the village Stampa. As for now the dam was used for the first and only time in 1987 when the Bergell region was subjected to a torrential rainstorm, twice as large as in 1927. The N-side of the dam is a popular place for climbing practice.

Along the road to Orden you immediately observe on the left a heavily deformed schist with spherical inclusions. This is part of the Austroalpine Margna nappe and the large spherical inclusions are feldspars floating in a micaceous matrix [#a2, 773.41/140.54]. Further along the road there are outcrops of less deformed "augengneiss" with large alkali feldspar crystals [#a3, 773.58/140.52]. The large crystals look like eyes, hence the German name "Augengneiss" (Fig. 4.2a). These gneisses were originally granite and intruded during the Carboniferous period ~300 million years ago into Pangaea. During the Alpine tectonic thrusts they were transformed into gneiss.

You will soon come to a fountain on the left. Make a short detour on an old road with cobblestones to the left (Fig. 4.2b). After about 100 m you will see on the right side an outcrop of Margna gneiss with beautiful polished surfaces and striations formed by the old Ice Age Forno glacier (Fig. 3.14b). Rub your fingers along the striations back and forth. In one direction it will be smooth, in the opposite more prickly. This tells you in which direction the glacier was moving (N). Then return to the fountain and continue along the main road. Most of the surface is glacial moraines, covered by meadows.

At the bridge of Orden (there are good outcrops along the Orlegna River ([a4], 774.00/140.15) the Engadine Line transverses, which separates the low grade metamorphic Maloja zone from the Bergell granite and its metamorphic shell. Gneiss has undergone strong semi-brittle deformation with fractures and associated kink-shaped structures. The Engadine Line has a left-lateral shift (the opposite part from where you stand has moved to the left), and the northern part is deeper (e.g. Schmid and Froitzheim 1993).

Fig. 4.2 **a** Outcrop of deformed Margna gneiss along the road to Orlegna with large alkali feldspar megacrysts (width 15 cm). **b** Old cobblestone road to the outcrop of glacial polish shown in Fig. 3.14b. **c** Cavloccio Lake from glacial times with a view south towards Muretto Pass. On the left is the Margna nappe (Africa) and on the right Monte del Forno with the contact zone of the Platta nappe and Bergell granite. You are standing on amphibolites that were at some time basalts of the Tethys oceanic crust. **d** Large prismatic andalusite crystals forming at low pressure–high temperature conditions; coin diameter is 18 mm. **e** Cavloccio amphibolite with large crystals of plagioclase feldspar, coin diameter is 30 mm. **f** Alluvium in the Orlegna river with Bergell granite (white) and a large boulder of granite with dark inclusions (xenoliths)

4.1 Easy Hikes

We follow the road and at a curve you observe a swampy meadow to your right ([#a5], 774.22/139.85). This is peat bog with humus-rich soil and interesting carnivorous plants such as Pinguicula and Drosera (Fig. 3.5d and Sect. 3.4.). South of the Engadine Line we enter metamorphic rocks with good outcrops on the right side of the road. First light-green tremolite-chlorite-albite schist (774.5/139.3) and soon after (774.33/139.95) dark-green amphibolite composed mainly of hornblende and plagioclase (Amphibolite metamorphic facies (Fig. 2.7). After passing a bridge (774.49/139.31) there is a large outcrop of strongly deformed pelitic schists that were originally sediments transformed to highgrade gneiss with minerals such as biotite, garnet (small spherical inclusions) and andalusite. Observe the small folds indicative of plastic deformation. These metamorphic rocks recrystallized during the intrusion of the Bergell granite in the Tertiary (~ 30 million years). Continue on the road to the Lake of Cavloc (Lägh da Cavloc) on your right and, before reaching a building with a restaurant, turn to the right on a small well-marked trail which makes a loop around the lake (Fig. 4.2c). There are beautiful pine forests (Pinus cembra, Arve, which grow at the highest forested elevations in the Alps). Immediately at the lake there are excellent outcrops of paragneiss with large prismatic andalusite crystals (Fig. 4.2d), typical of metasedimentary contact rocks forming at high temperatures and relatively low pressure (c.f. Fig. 3.11; [#a6], 774.48/139.08). There are no outcrops of Bergell granite along this trail but in the moraine west of the original glacial lake we see many boulders with the typical texture we described earlier and large alkali feldspar crystals (e.g. Figs. 3.3a and 3.6). This is similar to how Margna gneiss looked before it was deformed and recrystallized at low temperature conditions ([#a3], Fig. 4.2a). You continue on the trail around the lake which takes you back to the road.

East of the lake, near Alp da Cavloc ([#a7], 774.53/138.76) and also along the Orlegna River, there are outcrops of amphibolites with dark green hornblende and white plagioclase (Fig. 4.2e). They have a volcanic origin as basalts and were then transformed at high temperatures ("amphibolite facies", Fig. 2.7) during the intrusion of the Bergell granite. This location is extremely interesting for mineralogists because it is so far the only documented place where the two end members of the plagioclase solid solution albite ($NaAlSi_3O_8$) and anorthite ($CaAl_2Si_2O_8$) (Fig. 3.12) coexist (Fig. 3.10c shows a thin section of this sample in which albite (Ab) and anorthite (An) are marked; see Sect. 3.3).

From the Alp da Cavloc huts walk a bit further South following the trail towards Val Forno, pass a saddle and then turn left towards the Orlegna River ([#a8], 774.9/138.1). The river alluvium is a good place to look at the varieties of rocks, such as different types of granite, some with dark inclusions called xenoliths, schists, marble and amphibolite (Fig. 4.2f). This can be combined with a picnic.

On this hike (4.1.1) we now return to the car park either on the same path we came or we can take detours on marked trails. Most challenging is to ascend from Lägh da Cavloc via Pass dal Caval [#a9] and Motta Salacina to Bitabergh Lake (1–2 h). As you climb up you are in the old moraine of the Forno glacier with lots of Bergell granite. After the pass you are in metamorphic rocks, mainly amphibolite which is partially transformed to chlorite schist at a lower temperature. There are local moraines but no more Bergell granite.

A second option is to take a more or less horizontal trail from Lägh da Cavloc over Zocheta to Bitabergh Lake [#a10] (1 h). First you walk over amphibolite but after Zocheta you are on the moraine of the old Forno glacier composed dominantly of Bergell granite with large feldspar crystals.

From Bitabergh Lake take the trail back to Orlegna. You will pass a dam that was built in 1971 to prevent flooding after heavy rainstorms. Just across the dam, below a big palace-like mansion and near the parking lot (773.33/140.42) there are brown metasedimentary chlorite schists of the Margna nappe, containing chalcopyrite ($CuFeS_2$) and zircon ($ZrSiO_4$).

On this trip you have received a first impression of the glacial morphology (glacial polish, moraines, peat bog, lake), old rocks such as Margna gneiss, young transformed rocks such as amphibolite and andalusite schist, as well as boulders of Bergell granite in the river and in moraines.

4.1.2 Maloja: Torre Belvedere, Moraines, Peat Bogs and Glacial Mills (Map 2, waypoints #b1–#b3) (1–2 h)

Park in Maloja near the post bus stop [#b1]. South of the Post Office is the Segantini studio and house which is still used by descendants of Giovanni Segantini, a famous Italian painter. They organize tours: https://www.bregaglia.ch/en/segantini-trail. We take the Via Segantini for a hike to the west. The road ascends to Torre Belvedere with a magnificent panoramic view ([#b2], 773.17/141.39, Fig. 4.3a). However, this is not a medieval castle but was built in 1882–1903 by the Belgian Count Camille de Renessee and never finished. Today, the tower is used for temporary exhibitions. Count de Renessee wanted to develop Maloja, which was just a small village belonging to the municipality of Stampa, into an international resort. In 1884 he built the 300 room Maloja Palace Hotel in

Renaissance style. At the time it was the largest hotel in the Alps, overlooking Lake Sils, as an old postcard from 1890 displays (Fig. 4.3b).

From here we take a circular trail to seven glacial mills carved by the original Forno Glacier in Margna gneiss ([#b3], Fig. 3.14a; 773.06/141.81) and return to the parking lot. Along the trail you find peat bogs on meadows within the forest.

4.1.3 Val Maroz: Greenschists, Bündnerschiefer, Travertine, Serpentinite (Map 4, waypoints #c1–#c8) (4–6 h)

From a Casaccia parking lot ([#c1], 771.09/140.04) take the road or marked paths into Val Maroz (it is not accessible to cars). Wherever possible, one should use the newly excavated medieval path with old cobble stones as a shortcut (Fig. 4.1b, [#c2]). This path had already been used by the Romans to traverse the Alps via the Septimer Pass to the north. The Latin name "Septimer", suggests thet it is named after emperor Septimius, as the Julier Pass is named after Julius Caesar. Before turning into the mountain valley, you traverse light green rocks ([#c3], 770.33/140.56). They are weakly metamorphic albite-epidote-chlorite gneiss, also known as greenschist or "prasinite" (Fig. 3.4b) and belong to the Lizun unit of the Platta nappe. They have a mafic composition, similar to amphibolite but are of lower metamorphic grade. We will see them again and again on this trip.

After the steep ascent you come to 1790 m with a view on the flat and wide Val Maroz with many pastures for cattle. Here outcrops just before a gate show serpentinite (green) and, across an old bridge blueschists with riebeckite and stilpnomelane ([#c4], 769.88/140.58). A trail towards Septimer Pass branches off towards the north. We follow the main road to a bridge leading to Alp Maroz Dora and then choose the dirt road on the N-side of the river.

On the S-side of Val Maroz there is a range of mountains of the Lizun tectonic unit, belonging to the upper Pennine Platta nappe with S-dipping layers (Fig. 4.4a). At Piz Lizun (left), which can be reached by climbing up a slope of talus above Maroz Dora, there are basaltic-andesitic dikes (diabase) that indicate an episode of Tertiary volcanism. They are slightly older than the Bergell granite (35–40 My) and there was no retrograde transformation in an early phase of Alpine tectonics. Unlike dikes in Piz Grevasalvas and Piz Lagrev granite (Excursions 4.2.2 and 4.2.3, Fig. 3.3f), the minerals in the Lizun dikes contain unaltered crystals of plagioclase and dark crystals of hornblende. We can find the same dikes on the S-side of Piz Lizun and near Alp Furcela (excursion 4.2.5), where we will discuss them in more detail (e.g. Fig. 4.16a, b).

Fig. 4.3 **a** Torre Belvedere, Maloja view north towards Piz Grevasalvas (courtesy W. Hunkeler). **b** Palace Hotel with Torre Belvedere behind it and in the background the Bergell granite peaks on the left and Monte Gruf on the right (old postcard from 1890)

Fig. 4.4 **a** View on southwest side of Val Maroz with Piz Lizun (left) and Pizzi di Maroz (center). Extensive talus slopes. **b** River boulders with calcareous schist (Bündnerschiefer) (left) and greenschist (right) in Val Maroz. These are some of the major rock types in upper Pennine nappes. **c** Travertine in Val Maroz with calcified "waterfall". **d** Calcified moss (width 5 cm). **e** Octahedral magnetite crystals in chlorite schist (coin diameter 17.9 mm)

Here we stay in the valley on the north side and hike west towards Maroz Dent. On the right side (N) of Val Maroz you can see rock faces with grey calcareous schists. These were Jurassic limestones/slates, called "Bündnerschiefer", and are part of the Pennine Avers nappe. They are often interlayered with prasinites and we see good blocks along the way, particularly in river channels ([#c5], Fig. 4.4b). Occasionaly there are also blocks and boulders of Triassic marble. But be aware: on the dirt road you will also find small pieces of Bergell granite. They have nothing to do with Val Maroz but came from the gravel pit in Casaccia (Fig. 3.20d).

At coordinates 768.1/140.3 [#c6] there are springs originating on the northern slope with calcareous travertine sinter terraces (Fig. 4.4c) reminiscent of much larger terraces in Yellowstone Park (USA) or Pamukale (Turkey). Plants such as moss are converted into calcite (Fig. 4.4d). They are a testimony to the carbonate-rich rocks in this zone, where calcite dissolves and reprecipitates. Another good place for recent travertine is east of Lägh pit da la Duana but it is not so easy to access.

At Maroz Dent there are rockfalls that cover the whole valley. We traverse the main river Maira over a bridge and choose the path to the west towards Val Duana (not south towards Val da Cam!), first through a plane which is often partly flooded and with a good view on the landslide in the north (Fig. 3.16c; [#c7]).

Where a side stream comes down from the N, the Maira can be crossed over a bridge which we use and after 200 m explore pebbles in the stream. In chlorite schists we find magnificent crystals of octahedral magnetite (Fig. 4.4e; [#c8], 766.70/140.30). It is associated with serpentinites.

We return to Casaccia the same way we came. At the end of the hike take a stroll through the old village and look at some of the houses partially covered by the 1673 landslide with the old main entrance now part of the cellar (Sect. 3.5; Fig. 3.16b).

4.1.4 Albigna Hut: Bergell Granite, with Cable Car (Map 5, waypoints #d1-#d4) (4-8 h)

Outcrops of Bergell granite are not found on any road in Switzerland. You have to walk to an outcrop of this rock for at least two hours. As already Studer (1851) observed, granite can only be found in the summit part, while the valleys are composed of gneisses. This is an expression of the tectonic structure (Fig. 2.8). Figure 2.10c gives an overview of Val Albigna, with mountains and glaciers. Here all rocks are associated with the Tertiary Bergell intrusion.

The easiest way to Val Albigna is to take the EWZ cable car from Pranzaira (northeast of Vicosoprano [#d1], Fig. 4.5a) and get into the middle of the granite at the base of the dam with a large lake behind (Fig. 4.5b). It is the main part of the hydroelectric power system of EWZ (Elektrizitätswerke Zürich, Sect. 3.6.6) and was built in 1957. From the cable car station it is best to follow the old road on a loop with excellent outcrops of various types of granitic rocks [#d2]: Megacrystalline Bergell granite (granodiorite) with often parallel oriented alkali feldspar crystals. It is sometimes crosscut by aplite dikes, and occasionally coarser pegmatite dikes (Figs. 3.3e and 4.5c). Locally granite contains dark inclusions, rich in biotite and hornblende, so-called xenoliths (Fig. 4.5d). They represent fragments of contact rocks such as amphibolite that were incorporated in the granitic magma.

4.1 Easy Hikes

Fig. 4.5 **a** EWZ cable car from Pranzaira to the Albigna dam. **b** View on the dam and lake from the trail to the Capanna Albigna (SAC). Note lots of vegetation including trees at high altitude. **c** On the trail you pass through outcrops with Bergell granite (large feldspar phenocrysts) crosscut by an aplite dike (hammer is 50 cm long). **d** In the granite you often find dark inclusions (xenoliths). **e** Surface with glacial polish and striations just below the SAC hut. **f** If you choose to take the trail to walk down to Pranzaira, there are interesting rocks, including contact between granite and metamorphic rocks and a big fracture zone

You cross the dam with a good view of Val Albigna with imposing peaks and a glacier in the background. Climb on a path to the Capanna Albigna (SAC) and along the way observe the flora with blueberries, junipers and especially young larch trees and birches (Fig. 4.5b). These are new at this elevation (2300 m) and are probably associated with climate change over the last 50 years. Before reaching the hut there are outcrops of granite with glacial polish and linear striations ([#d3], Fig. 4.5e).

Return to the cable car station. On the way down you may choose not to take the cable car but instead follow a marked trail to Pranzaira. Before the steep descent you pass an impressive fracture, reminding us that many rocks in the Bergell are also subject to late brittle deformation (Fig. 4.5f). The trail starts in granite, then transverses tonalite (hornblende-biotite-plagioclase; Fig. 3.3c) and soon you reach the contact with high-grade metamorphic rocks [#d4]. At the bottom turn east and pass the Albigna river over a bridge (770.28/136.31) and from there return to Pranzaira.

4.1.5 Maira-Vicosoprano: Diversity of Bergell Rocks in the River (Map 5, waypoints #e1–#e2) (1–2 h)

With excursions such as Cavloc, Forno, Albigna, Piz Lunghin, Val Maroz and Val da la Duana, you may have become familiar with the variety of Bergell rocks. A good place to explore the diversity is in river alluvium. We mentioned the Orlegna river south of Alp da Cavloc (excursion 4.1.1, [#a 8]) and will suggest other river locations such as the Mera near Prosto (excursion 4.1.6, 752.12/132.48) and the Torrente Codera at Novate Mezzola- Mezzolpiano (excursion 4.2.9, 755.96/120.66). These riverbeds provide a wide average of rock types, not only over rock outcrops but also rocks contained in moraines. Pebbles and blocks of cohesive rocks such as granite dominate over schists and serpentinite which easily erode to sand. In the whole Bergell region Bergell granite is a dominant component in main rivers. But alluvium is also subject to rapid changes. There was an excellent outcrop in the Maira at Pranzaira near the cable car station (769.54/136.69) with a wide variety of rocks from the upper Bergell Valley region (Fig. 17b), including a big block of Bergell granite with a dark xenolithic inclusion in the shape of a question mark which was reminding geologists for over 50 years that many issues about Bergell rocks and tectonic processes are still unresolved (Fig. 4.6a).

On July 24, 2023 while this book was in production, a large landslide from Valun del Largh, causing little damage to structures but covered the whole Maira river from Roivan to Vicosoprano largely with Bergell granite and at the same time extinguished the fish population. Thus we are shifting the rock review further southwest.

Park the car near the bridge of the road going to Rotticcio ([#e1], 768.65/136.07), then follow an unpaved track to the west, cross a meadow and a small forest section and descend to the Maira river ([#e2], 768.36/136.08). In the alluvium of the Maira many different rocks come together and it is an ideal place to start a small rock collection. Bergell granite dominates among the pebbles (Fig. 3.3a). There are different types of granite, mostly with large potash feldspars, but also variants with dark inclusions as we have observed in Val Forno

Fig. 4.6 River debris in the Maira near Pranzaira. **a** Large boulder of Bergell granite with question mark. **b** Various pebbles of Bergell rocks. Top left to right: Tonalite, Bergell granite with large alkali feldspar, aplite dike with gneiss (left), fine-grained Bergell granite; center: epidote schist, serpentinite, quartzite, marble with scratch, calcsilicate with garnets near contact; bottom: amphibolite, two greenschists, biotite-muscovite gneiss, micaschist. Pebbles are about 8 cm wide

and Val Albigna. Then there are metamorphic rocks such as gneiss, amphibolite, greenschist, marble and quartzite. They are less stable than granite and more easily degraded into sand in the river. They originate from a much wider range than Bergell granite such as nappes on the N-side of the Bergell and from the granite contact zone. Figure 4.6b is a collection of fifteen typical Bergell rocks that we can assemble in half an hour. White marble can be scratched with a knife or a hammer, white quartz scratches the knife. Compare the rocks with those displayed in Figs. 3.3 and 3.4.

After alluvium and Bergell rocks you may want to visit a nearby lime kiln north of Crot in the forest, less than a kilometer south of Pranzaira near the main road (769.18/136.32, see also Sect. 3.6.2).

4.1.6 Piuro-Borgonuovo, Palazzo Vertemate-Franchi, Explore Lavez and Glacial Features in a Walk from Prosto (Map 3, waypoints #f1–#f8) (2–4 h)

In this section we describe short excursions between Piuro and Chiavenna. You can do all this in one day or spread it over different days and combine it with other projects.

We start in Piuro-Borgonuovo west of Villa di Chiavenna. Here was once the large town of Plurs/Piuro which was destroyed in 1618 (Fig. 3.15a and Sect. 3.5). There are some ruins on the south side of the valley and a single lavez quarry still operates (Fig. 3.18f, 754.23/132.06). More attractive is the waterfall on the northwest side, Aquafraggia, below the village of Savogno ([#f1], Fig. 4.7a).

Next we continue on the valley road to Piuro-Prosto and turn north on Via Palazzo Vertemate to Palazzo Vertemate-Franchi [#f2]. This masterpiece of the Renaissance survived the landslide of Plurs and is now a museum (Fig. 4.7b). In the garden you will find fountains and irrigation canals made with lavez (Fig. 4.7c). In the courtyard of Palazzo Vertemate you can observe round paving stones of lavez, the remains of pot making (Fig. 4.7d, compare with Fig. 3.18e).

We have described lavez rocks in Sect. 3.6.1. Lavez (Laveggio, Pietra Ollare, soapstone) consisting mainly of talc and olivine was the main industry in old Plurs-Piuro. Today there is still a workshop in Prosto where lavez pots are prepared and is worth a visit (Contact Roberto Luchinetti https://www.pietraollare.com/chi-siamo/) and park at the church on the other side of the Maira ([#f3], 752.60/132.35). Larger productions still exist in Val Malenco.

4.1 Easy Hikes

◀**Fig. 4.7 a** Large waterfall Aqua Fraggia in Piuro-Borgonovo. **b–d** Palazzo Vertemate-Franchi in Piuro-Prosto, the only palace that survived the 1618 landslide. **c** Lavez basins and channels are used for irrigation and **d** remains of pots are used for cobble stones (c.f. Fig. 3.18e). **e–h** Roundtrip from Prosto starting south of the river near the church. **e** Impressive glacial polish with striations and prehistoric incisions such as a snake (enlarged in upper left). **f** Very large glacial mill (Marmitte dei Giganti) 4 m in diameter. **g** Boulder of Bergell granite on a polished glacial surface. **h** Remains of a lavez quarry. The round indentations reveal where blocks were removed for pot production (c.f. Fig. 3.18d)

From here we take the path towards Uschione, up the hill. The trail crosses outcrops of glacial polish with prehistoric incisions such as two snakes and other symbols ([#f4], Fig. 4.7e, 752.66/132.20). We follow the path "Marmitte dei Giganti" which branches off the Uschione path and comes to a saddle (Passo Capiola). Just above the turnoff there are cliffs on the N-side with beautiful small glacial mills. Then we climb north to the ridge (Pt. 522) with a panoramic view (Sasso Dragone). Look at glacier relics with glacial mills (cliffs on W-side), glacial polish and erratic blocks of Bergell granite. On the descent on the ridge to the west you will find a really gigantic glacial mill (Marmitte dei Giganti, over 4m in diameter in Tambo gneiss, [#f5], Fig. 4.7f, 752.27/132.33).

We continue on the path and pass a couple of old stone buildings. Then we choose a smaller trail to the left to stay on the ridge. It takes us to glacial surfaces on Tambo gneiss with a large round block of Bergell granite that made the 30–40 km long journey from Forno and Albigna ([#f6], Fig. 4.7g). Relics of Bergell granite are found much further away, at the S-end of Lake Como, about 100 km from Maloja Pass. Also, as mentioned earlier (Sect. 3.4), Bergell granodiorite as well as tonalite were already deposited long before the Ice Age in late Oligocene-early Miocene conglomerates also south of Como (c.f. Fig. 3.14d, Di Capua et al. 2015).

We follow the trail to the west and on a hill you have a nice view overlooking the town of Chiavenna [#f7]. Then we descend to the north and come to outcrops of lavez with old mines where typical round relics are preserved ([#f8], Fig. 4.7h, compare with Fig. 3.18d and review Sect. 3.6.1). We continue into the valley at the eastern end of Chiavenna and return along the Mera River to Prosto, with many old romantic Crotti (little caves) on the way. On the way there are several stairs descending to the Mera river, where you may want to continue with your pebble collection (Excursion 4.1.6). Even though outcrops are far away, there is a lot of Bergell granite in the river alluvium.

4.1.7 Chiavenna, Outcrops of Tambo Granite in the Liro River, Cimaganda Landslide (Map 3, waypoints #g1–#g7) (2–3 h)

Chiavenna is currently the largest town in the region with many old buildings. Spend a couple of hours exploring the city and links to geology. For a geological map consult Montrasio and Sciesa (1988) and Map 5. Convenient parking is near the cemetery at the southeast end of the city ([#g1], 751.64/132.00). Here is also the museum with history and minerals, and you can take a tour of lavez quarries (Parco Archeologico Botanico del Paradiso e Museo Archeologico della Valchiavenna). A visit is recommended.

Near the parking area is Collegiata di San Lorenzo (Fig. 4.8a) and when you enter, at the second door to the left, there is a small chapel with a large Romanesque baptismal font from 1156 ([#g2], Fig. 4.8b). It is carved from lavez (pietra ollare). Then we pass through old streets (Fig. 4.8c) to Piazza Rodolfo Pestalozzi (no relation to the Swiss teacher Johann Heinrich Pestalozzi). Here is an impressive lavez fountain ([#g3], Fig. 4.8d). You can also find lavez cores in walls of old buildings (Fig. 4.8e) as you may have seen among the cobble stones at the Palazzo Vertemate-Franchi (Fig. 4.7d).

From Chiavenna we drive north towards the Splügen Pass. Soon after we leave Chiavenna and before San Giacomo Filippo we park on the side of the highway ([#g4], 749.31/132.32) and descend to the Liro River on the left (Fig. 4.8f). Here we find beautiful outcrops of granite ([#g5], 749.18/132.32, Fig. 4.8g). This Tambo granite resembles Bergell granite. However, it is an old granite of Variscan age (~ 300 My) of the Tambo nappe, corresponding to gneiss at the Soglio and Promontogno quarries where it has been transformed through tectonic deformation (Fig. 3.20a, b). At the Liro outcrop granite was also locally transformed to gneiss (Fig. 4.8h) and you see both, side by side.

Next we continue on the Splügen road to Cimaganda with parking at the beginning or the end of the village ([#g6, not on map], 747.76/137.24). This town was largely destroyed by a landslide in 1533, but was rebuilt in the middle of the landslide between large blocks of Tambo gneiss surrounding buildings (Fig. 4.9b). There is a good view of the landslide and the village from the western slope which can be reached over a small bridge at the bottom of the village ([#g7, not on map], 747.52/137.51). Also on this side you see buildings destroyed by rockfall.

From here you have a good view of the steep E-side (Fig. 4.9a). Cimaganda is on the right, relics of the 1533 landslide in the center. Then there was a more recent rockfall on the left below Bondeno in 2012 which interrupted the

84 4 Geological Excursions in the Bergell Alps

◄**Fig. 4.8** a–e Some pictures of Chiavenna which became the center after the devastation of Plurs/Piuro. **a** Church San Lorenzo with **b** a large lavez baptismal font from 1156 (courtesy Caritas Ticino). **c** Road in old town with lavez block on right. **d** Large lavez fountain on Piazza Pestalozzi. **e** Also in Chiavenna lavez cores were used to rebuild structures. **f–h** Outcrops of Tambo gneiss at the Liro River between Chiavenna and San Giacomo Filippo. **g** Granitic variety with megacrysts (compass length is 20 cm). **h** Deformed variety of augengneiss with foliation, similar to Tambo gneiss in the Bergell

Fig. 4.9 Rockfalls in Cimaganda along road to Splügen. **a** Several rockfalls east of Cimaganda. Left is a rockfall 2012 that covered the valley highway, center the large rockfall 1533 that partially destroyed the village and far right a recent rockfall (2018) above the Santuario di Gallivaggio. **b** A large block from 1533 in the center of the village. **c** A view West. Also here many rockfalls occurred

main valley highway. In 2018, another landslide on the right (S) at Santuario di Gallivaggio miraculously stopped just above the old church. Rockfalls are also common on the W-side (Fig. 4.9c) which is equally covered by large blocks. Obviously the rocks are very unstable on both sides of this valley, with its large cliffs. Cimaganda gives you an impression of the most dangerous natural disasters in the Bergell Alps: landslides and rockfalls (remember Figs. 3.15, 3.16 and 3.17). They are more common and frequent than earthquakes in California.

4.2 Longer Excursions (1–2 days)

4.2.1 Bergell Granite and Contact Zone, Val Forno Glacier, Forno Hut, Monte del Forno, Pillow Structures, Muretto Pass (Map 6, waypoints #h1–#h10) (2 days)

Day 1: Maloja-Cavloccio-Val Forno. Andalusite schists and amphibolites around Cavloc, Forno glacier. Bergell granite, xenolith swarms. Overnight stay at Forno Hut, SAC.

First follow the notes for "Easy Hikes, 4.1.1" on Map 1 to Alp da Cavloc [#a]. The outcrop at the lake with large prismatic andalusite crystals is spectacular ([#a6], Fig. 4.2d). Also examine amphibolite ([#a8], Figs. 3.10c and 4.2e).

Then we continue hiking towards Plan Canin and Val Forno (Map 6). You cross a saddle and soon after you step on the trail on a plate full of prismatic andalusite crystals (Fig. 4.10a). Continue until you reach a trail marker where one path goes towards Muretto Pass and the other one to Forno ([#h1], 775.11/ 137.56). This is an interesting place with outcrops of glacier-polished surfaces and heavily deformed metamorphic rocks. These rocks show folds formed during horizontal compression (Fig. 4.10b) in a kink-like manner. Another feature known as "boudinage" in structural geology, first described in slates of the Belgian Ardennes, can be viewed (Fig. 4.10c). A vertical layer of stiff quartz in a more ductile micaceous matrix is pulled apart into fragments with the matrix filling gaps. The quartz fragments now form "a string of sausages".

In Plan Canin there is a small reservoir from which the water flows in a tunnel first to Murtaira and then, combining with water from the Albigna Lake to the EWZ electricity plant in Löbbia (rocks in the tunnels, documenting the contact between Bergell granite and its metamorphic surroundings have been studied in detail by Weibel and Locher (1964). Near the house there is a display which illustrates the retreat of the Forno glacier over 100 years. In Sect. 3.4 we describe

4.2 Longer Excursions (1–2 days)

Fig. 4.10 Fieldtrip to Forno, Day 1. **a** After Alp Cavloc you pass a small saddle and step over a block with large prismatic andalusite crystals. Knife is 9 cm long. **b** Strongly folded gneiss at trail intersection Plan Canin(width 1 m). **c** Same place with boudinage where a strong quartz layer is pulled apart by a softer micaceous layer (width 1 m). **d** Contact between amphibolite (right) and granite in lower Val Forno (west side). **e** Climbing up from the Forno glacier on the E-side provides an overview. **f** Locality with outcrops of mafic xenolith swarms in granite. **g** Outcrop of fine-grained granite with orbicular structures (width 60 cm)

that 150 years ago, the Forno Glacier reached Plan Canin (Fig. 3.13e). Today it can be found ~4 km further south (Fig. 3.13d); on average it retreats about 30 m per year. In the Roman "Climate Optimum" there were probably no glaciers in the Bergell (Fig. 3.13b). They extended during the following cold period (Little Ice Age). Today's rapid glacier retreat is certainly influenced by anthropogenic climate changes.

From here we take the trail into Val Forno. Soon after the reservoir we get a good view on the contact zone with amphibolite and granite ([#h2], 774.80/ 137.00, Fig. 4.10d). The large alkali feldspar crystals (orthoclase) are typical of Bergell granite (Figs. 3.3a and 3.6). On the trail we pass through a rockfall with two different varieties of granite: The northern-most granite next to the amphibolite is a slightly orange-colored variety caused by partial alteration of the quenched melt next to the cold iron-rich environment. Biotite is partially altered to chlorite and plagioclase contains microscopic inclusions of iron-oxide. A subsequent intrusion a bit further south is grey and much more homogeneous.

As you reach a flat area at 2200m the trail crosses the river over a bridge to the east and you traverse two moraines on the east side. The first (W) consists mainly of granite with typical large alkali feldspar crystals, the second (E) is mainly of marble from Cima di Vazzeda in the rear of Forno Valley [#h3], 774.30/134.46). In the marble there are excellent samples of the minerals tremolite, diopside, grossular garnet, vesuvianite and clinohumite, which crystallized when original limestones were transformed as the hot granitic melt intruded. There are also many red-brown pebbles of metamorphic schist that contain ore minerals, here mainly pyrite which you can identify by hitting it with a hammer and tasting the sulfuric smell. Some ores in the Forno region were processed at Plan Canin where trees could be lumbered and used for smelting (hence the name "Forno", see Sect. 3.6.4).

Approximately at 774.75/134.20 we leave the main marked trail to the Forno hut and take a side path towards NE. At 2600 m (774.94/134.33) there is a beautiful view on upper Val Forno (Fig. 4.10e). Here we find outcrops of granite with swarms of dark inclusions, rich in hornblende and biotite, first described by Gansser and Gyr (1964) ([#h4], Fig. 4.10f). These so-called xenoliths (xeno Greek for "foreign" and lithos for rock) are remnants of contact rocks such as amphibolite that were broken up into fragments, "float" in granite and were transformed. There are similar swarms of xenoliths in Val Bondasca, but there they are much more deformed and aligned (excursion 4.2.9).

At 2700 m we reach a more or less horizontal trail northwards to a small lake. Here we find another variety of Bergell granite, lighter (less biotite), without phenocrysts and often with round inclusions of dark hornblende, so-called "orbicular granite". The orbicules have a spherical shape and indicate growth around a mafic nucleus ([#h5], Fig. 4.10g, 775.04/134.55).

Now return on the trail south and spend the night in the Capanna del Forno (SAC) ([#h6], 774.83/133.70).

Day 2: Sella del Forno (variant Monte del Forno), contact zone with amphibolite, pillow structures, Muretto Pass with Margna nappe. Maloja.
On the second day we climb on a blue-white marked trail to the Sella del Forno. A short detour to the south leads to the foot of Monte Rosso ([#h7], 775.91/133.02), with another occurrence of orbicular granite, and good examples of Bergell granite, pegmatite (coarse-grained dike) and aplite (fine-grained dike). These veins crystallized from the water-rich residual melt of the magma and are often zoned between edge and center (Fig. 4.11a). Along the contact the granite is often deformed with aligned feldspar megacrysts (Fig. 4.11b). Locally granite intrudes amphibolite like a dike (Fig. 4.11c). From here you also have a magnificent view to the north with Monte del Forno and below it a system of pegmatite dikes that have intruded into the amphibolite of the contact zone (Fig. 4.11d, e).

From the Sella del Forno we climb on a trail along the ridge northwards to point 2942 ([#h8], 776.16/133.94), mainly in amphibolite. One variant is a detour to Monte del Forno (3213 m) further along the ridge, technically not very difficult, with blue and white markers, but a rope is recommended, particularly if there are slippery or icy surfaces. From Monte del Forno ([#h9], 775.99/134.44) enjoy a beautiful view in all directions: towards southwest the mountains of the Bergell massif, to the W, behind Val Bregaglia, is the stack of Pennine nappes (old Europe), to the east the Austroalpine Margna nappe (Africa) with Piz da la Margna, Piz Fedoz, and behind it the Sella and Bernina nappes. Then descend from the summit back to point 2942.

The descent from Pt. 2942 towards northeast leads on a trail through contact metamorphic schists and amphibolites. At 2700 m the trail extends horizontally. Amphibolites, now metamorphic rocks composed largely of hornblende and plagioclase, often have "pillow structures" (e.g. [#h10], 776.11/135.23, Fig. 4.11f), reminding us that these rocks were at one time oceanic basalts and the melt was quenched and solidified in cold water. Today, pillows form in places like the southside of Hawaii Island. These amphibolites with pillows are part of a mafic/ultramafic ophiolite zone with serpentinites in Valmalenco in the south, amphibolites near the Forno contact, greenschists in the Lizun complex and serpentinites again in the north, belonging to the Platta nappe (green part in Fig. 2.4; e.g. Bernoulli and Weissert 1985).

Fig. 4.11 Fieldtrip to Forno, Day 2. **a–c** Contact zone south of Sella del Forno. **a** Bergell granite with some xenoliths (Ice ax pick is 35 cm). **b** Aplie dikes Applite dikes and Bergell granite with highly deformed and aligned phenocrysts near the contact (width 50 cm). **c** Granite intruding into amphibolite (width 1 m). **d** View north with Monte del Forno (left) and Piz da la Margna in the background (center). **e** Detail of amphibolite transected by pegmatite dikes. **f** Pillow structures in amphibolite, reminiscent of a submarine origin, above Muretto Pass (hammer length 50 cm). **g** Overview with Muretto Pass, Austroalpine Margna nappe (left, E, "Africa") and the Forno contact zone (right, W, "Tethys ocean")

Amphibolites are transformed parts of oceanic crust in the Mediterranean Tethys Ocean, with parts of peridotitic upper mantle. Associated with the pillow basalts are deposits of rhodonite schists. Rhodonite ($MnSiO_3$) is a red manganese mineral, typical of oceanic sediments. In some localities there are small ore deposits that were briefly exploited for Cu and Mn (e.g. Maurizio 1972; Ferrario and Montrasio 1976; Peretti and Köppel 1986, see Sect. 3.6.4). There are also good outcrops with prismatic andalusite crystals in pelitic schists, similar to the ones observed at Lägh da Cavloc (#a6).

We descend to the Muretto Pass and on a good trail return to Plan Canin and back to Maloja. On the way back have another look southeast where you see mountains of the Austroalpine Margna nappe which lies above the Bergell granite (Fig. 4.11g). This nappe is "Africa" which was pushed over the oceanic/metasedimentary Tethys units and the Pennine nappes ("Europe").

4.2.2 Maloja, Pass Lunghin, Piz Lunghin: Nappes, Serpentinite, Triassic Marble (Map 7, waypoints #i1–#i5) (5–8 h)

We park in Cad Lägh east of Maloja, the earlier in the morning the better because parking is limited ([#i1], 774.21/142.38). Then we climb on the trail over Plan di Zoch to the Lägh dal Lunghin. In the ascent we first walk through old granitic gneiss (augengneiss with large alkali feldspars) of the Austroalpine Margna nappe [#i2], originally intruded ~320 My ago. At 2450 m one traverses marbles and quartzites [#i3], originally Triassic sediments deposited on granitic basement at ~240 My. Higher up there is an excellent view towards south with Piz da la Margna (Austroalpine) on the left, Monte Disgrazia in the back, composed of Malenco serpentinite, Forno amphibolites (Tethys Ocean) and Bergell granite on the right (Fig. 4.12a). From the W-side of the lake, it is best not to follow the path, but cross the stream to the north and hike on the moraine ridge or closer to the cliffs on the N-side. There is a greater variety of rocks: serpentinite (Pennine Platta nappe), Austroalpine Grevasalvas granite (part of the Err granodiorite), locally crisscrossed by old diabase dikes, banded and folded Jurassic calcareous schists (Austroalpine Err nappe).

At Lunghin Pass (Fig. 4.12b, point 2645 [#i4], 771.03/142.67) three drainage areas meet: on one side water flows northwest (Eva dal Lunghin, Bivio, Rhine, North Sea), on another southwest (Alpascela, Bergell, Maira, Po, Adriatic Sea) and on a third side of this "triple point" it flows northeast (Lägh dal Lunghin, Maloja, Engadine, Inn, Danube, Black Sea).

Fig. 4.12 a–d Excursion to Piz Lunghin. **a** Above Lägh dal Lunghin view south with Piz da la Margna (left), Monte Disgrazia (to the right of it in the background) and Val Forno with Forno glacier. **b** Triple point at Pass Lunghin where drainages to N-Sea (N), Black Sea (E) and Adriatic Sea (S) meet. **c** Climbing towards P. Lunghin over serpentinite (gray-green) and view on the Avers Alps (N). The peak in the foreground is Piz Grevasalvas. **d** View from Piz Lunghin summit into the Engadine with Lake Sils, Piz Lagrev (left) and the Bernina massif (right)

At the pass there are good outcrops of serpentinite, a magnesium-rich silicate rock that originated in the sub-oceanic upper mantle (olivine peridotite), was then pushed into the crust and transformed into serpentinite through reactions with water (Fig. 3.4d). The rock sepentinite consists mainly of the mineral serpentine. There are three polymorphs of serpentine minerals with similar composition but different structures: antigorite, lizardite and chrysotile (Table 3.2). Serpentinite is often blocky and has polished surfaces between blocks. The serpentinite zone of the Platta nappe stretches from Arosa (N) over Lunghin towards Val Malenco (S) (Fig. 2.4, green, e.g. Bernoulli and Weissert 1985).

From the pass, a blue-white marked path leads mainly over serpentinites (Fig. 4.12c) to the Piz Lunghin summit ([#i5], 771.57/142.34, 2780 m), one of the most beautiful panoramic mountains, with a view of the Upper Engadine (Fig. 4.12d) and the Bergell. From here you get an overview of geology: the Tertiary Bergeller granite massif in the south with Monte del Forno and Monte Disgrazia in the background (Fig. 4.12a); and to the west the stacks of Pennine nappes: Tambo nappe and Suretta nappe with Piz Duan.

In the wide U-shaped valley of the Engadine with lakes Sils and Silvaplana, the morphology created by the glacial erosion during the ice age has largely been preserved. In the Bergell, on the other hand, the valley is now bordered by steep cliffs with talus debris and the glacial morphology has been modified by rockfalls and landslides.

Descend to Cad Lägh, mainly through gneiss of the Margna nappe.

4.2.3 Plaun da Lej-Grevasalvas-Plaun Grand, Fuorcla Grevasalvas, Plaun dal Sel, Blaunca: Higher Nappes, Radiolarite, Old Granites (Map 7 waypoints #j1–#j10) (6–8 h)

First, a brief overview of plate tectonics (consult the review Chap. 2, Fig. 2.2), with the giant continent Pangaea (300 My) splitting up in the Triassic (200 My) into Gondwana in the South (Africa–south America–Australia) and Laurasia in the north (Asia–Europe–north America). Between south Europe and north Africa the Tethys Ocean is formed. In the Cretaceous period (145–66 My, Fig. 2.1), northern fragments of Gondwana (Africa) moved north and collided with Europe with the oceanic Tethys crust in between. This collision led to the formation of the Alps with a stack of European (Pennine) nappes at depth, oceanic crust in between (Platta nappe) and African (Austroalpine) nappes on top (Fig. 2.3).

The topic of Excursion 4.2.3 is to explore the Austroalpine nappes (Margna, Err) and their relationship to the Pennine Platta nappe. The tectonic processes are three times older than the intrusion of Bergell granite (30 My). We will observe metamorphic rocks such as marble, greenschist, blueschist, chert, serpentinite and ancient granites of the African crust. Figure 2.8b gives a tectonic overview and Fig. 4.13a displays the region we are going to explore.

We start in Plaun da Lej at Lake Sils ([#j1], 775.86/143.63). But first we take a short detour to an old dolomite quarry at Sasc da Corn ([#j2], 775.87/143.39) with an old lime kiln (Fig. 3.18a). These dolomites are Triassic sediments of the Margna nappe.

Then we follow the road to Grevasalvas with outcrops of metamorphic slate and serpentinite. In the meadows under Grevasalvas and before crossing the

Fig. 4.13 **a** View north with Piz Lagrev and Lake Sils, with stack of nappes Margna, Platta, Err-Bernina. **b** Sample of blueschist on road below Grevasalvas with prismatic stilpnomelane (brown) and riebeckite (blue) (width 4 cm). **c** Minerals are best recognized in thin sections, particularly blue riebeckite (width 5 mm). **d** Serpentinite along the trail below Plaun Grand (hammer length 50 cm). **e** View west with Pennine Platta nappe squished between Austroalpine Margna and Err nappes. **f** Red chert from Plaun dal Sel above Blaunca (coin diameter 19.1 mm). **g** Chert contains well-preserved radiolarian skeletons observed with an optical microscope (width is 1 mm)

Lavatera River there are interesting outcrops of blueschist with dark-blue riebeckite needles and stilpnomelane rosettes (Fig. 4.13b, [#j3], 775.35/143.85). The microstructure is best displayed in a thin section (Fig. 4.13c). This association is typical of metamorphism at low temperature and relatively high pressure (blueschist facies, Fig. 2.7).

On a marked path we climb from Grevasalvas to the north, first again through serpentinite. At the junction Point 2011 we make a small detour to the east to the debris field of the Ova da la Roda ([#j4], 775.61/144.39), where blocks of Err granite can be seen covering the southern slopes of Piz Lagrev (Fig. 4.13a). Like the Maloja augengneiss of the Margna nappe (Excursion 4.2.2), this Err/Lagrev granite is also old (about 330 million years) but in contrast to Margna gneiss the Err granite has not been converted into gneiss by deformation and metamorphism. The greenish color comes from the decomposition of feldspars to saussurite at low-grade hydrothermal conditions. Saussurite is a fine-grained mixture of zoisite, albite, sericite and epidote. The same granite of the Austroalpine Err nappe can be found at Piz d'Err, Piz Julier, Julier Pass, Piz Lagrev and Piz Grevasalvas.

We return to the Point 2011 and now take the trail northwest towards Plaun Grand. The path crosses a large landslide (Gianda). At 2280 m it traverses a serpentinite zone ([#j5], 774.30/144.24, Fig. 4.13e). Serpentinite is a magnesium-rich ultra-mafic rock with green colors that can be easily identified (Fig. 3.4d). It is formed when olivine rocks such as peridotite from the Earth's mantle reach higher elevations during upwelling and olivine reacts with water to form serpentine. On the excursion to Piz Lunghin the same zone was observed. The serpentinite rocks belong to the Platta nappe and thus to the uppermost part of the Pennine nappes.

Above Plaun Grand you can see cliffs with granitic gneiss on the right, which belong to the Err nappe (Fig. 4.13e). In the valley is the serpentinite of the Platta nappe and to the left are metamorphic sediments of the Margna nappe, including marbles. Thus the Pennine Platta nappe ("Tethys Ocean") is juxtaposed between two Austroalpine nappes (Err and Margna) during the various episodes of compression and extension. As the classic profiles of Cornelius (1950) show (Fig. 2.9) and already indicated by a sketch by Studer (1851) (Fig. 1.2), the structure in these higher Alpine nappes is complicated with different phases of thrusting. For more details see e.g. Handy et al. (1996).

We follow the path to the north towards Fuorcla Grevasalvas and cross first augengneiss, then a marble layer. Under Piz d'Emmat Dadaint we find red siliceous chert in debris ([#j6], 774.46/145.33). These are deep ocean sediments and we will discuss them later in more detail. We continue to Fuorcla Grevasalvas (2688 m, [#j7], 774.75/145.56). Now we are in the Err granite (northeast side) and towards the north we have a view on Piz Julier composed of the same granite. In the granite below Piz Lagrev there are occasionally dark dikes of diabase, which testify that at a later time, when the granite had already solidified, liquid rock intruded along fissures (Fig. 3.3f). But the diabase dikes are old and original minerals have been transformed during Alpine metamorphism.

On a clear day, a detour to Piz d'Emmat Dadaint ([#j8], 774.04/145.81, 2927 m) is worthwhile. From Fuorcla Grevasalvas you can simply follow the ridge to the west, first through old gabbroic rocks with hornblende which compose also Piz Materdell and belong to the Err nappe. From Piz d'Emmat Dadaint you have a great view in all directions, particularly to the north with a view on the Pennine nappes Avers and Platta in the Oberhalbstein with Piz Platta, and to the west with nearby Piz Materdell and Piz Grevasalvas.

On the descent you will cross red radiolarian cherts and soon reach the trail on which you came up. Return to the serpentinite outcrop [#j5] and from there follow the serpentinite zone along a diagonal path towards southwest to Plaun dal Sel ([#j9], 774.37/143.82). Here we again find red chert, associated with serpentinite. In thin sections the chert (Fig. 4.13f) displays well preserved spherical skeletal fossils, radiolaria, that have survived the Alpine metamorphism (Fig. 4.13g).

We reach Blaunca on a trail and pass through Margna marble. Just across the river Ova dal Mulin in the meadow is an old lime kiln worth exploring. For continuing your descent there are two variants: Either use the road to Grevasalvas and Plaun da Lej or, more interesting, take a trail from Blaunca to the southeast towards a ridge of Motta da Blaunca, then descending towards Splüga but before entering Splüga head east along the S-side of the ridge. Part of this is marble. The path leads through a valley with occasional peat bogs from post-glacial times with a specific flora including the carnivorous plants Drosera and Pinguicula ([#j10], 775.42/143.40) as described earlier (Fig. 3.5d). The trail meets the road to Grevasalvas and back to Plaun da Lej.

4.2.4 Casaccia-Val Maroz-Val da la Duana-Piz Duan-Cadrin-Löbbia-Soglio: Pennine Nappes (Avers, Suretta, Tambo) (Map 4, waypoints #k1–#k10) (2 days)

This is a long hike! It is easier to do it in two days with camping at Lägh da la Duana. Park your car in Casaccia and follow instructions described for excursion "Easy Hikes 4.1.3, #c1–c6". On this long tour you probably won't have time to look for minerals [#c7]. While you are hiking up through Val Maroz

4.2 Longer Excursions (1–2 days)

look at greenschists and calcareous schists (Bündnerschiefer) of the Avers nappe (Fig. 4.4b) and travertine (Fig. 4.4c, d).

At the bridge of Maroz Dent follow the path to the east and ascend to Lägh da la Duana (2466 m). First you traverse Suretta augengneiss and towards the top you can see quartzite and marble. On the shores of the lake there are wonderful marble outcrops with karst structures created by dissolution over centuries ([#k1], 765.38/139.68, Fig. 4.14a). Marble is plastically strongly deformed with folds (Fig. 3.4e). In the meadows behind you can find Edelweiss (*Leontopodium alpinum*) which is always associated with carbonate rocks (Fig. 4.14b). This lake does not drain through a stream but the river disappears in a hole, also caused by dissolution of the calcareous rocks, passing through a tunnel and reemerging a kilometer further down.

If time permits, a detour to Piz Duan is suggested (3135 m, Fig. 4.14c). It is one of the best panoramic mountains in the Bergell, technically not difficult but a rope is recommended. It will take 2–3 hours ([#k2], 765.01/138.33). You follow tracks, sometimes in snow, ice or gravel. Rocks at the bottom are quartzite and marble of the Pennine Suretta nappe and grey calcareous schists (Bündnerschiefer) of the Pennine Avers nappe towards the top. From the summit you can see the contrast between the rugged cliffs of the Bergell massif in the south (Pizzo Badile, Pizzo Cengalo, Fig. 4.14d) and the wide meadows in the north–east. On a clear day you can see Monte Rosa in the far west.

We return to Lägh da la Duana (short cuts from Piz Duan to west or south are possible but technically more difficult). Then we camp and the next day follow the trail to upper Val da la Duana to the west. At 2520 m you cross Triassic dolomite marbles above quartzites ([#k3], 764.23/139.19), the same ones you saw at the lake. Have a look back at Piz Duan, composed of Jurassic Bündnerschiefer (grey) of the Avers nappe on top and of Triassic marbles and quartzites (bright colors) and older metasedimentary schists below, resting on Suretta orthogneisses (right side) (Fig. 4.15a).

North of a small lake at 2575 m, Lägh pit da la Duana, there are outcrops of the pelitic schists (paragneiss), not only with garnet and chlorite but also chloritoid, a relatively rare mineral in greenschist-facies schists (Fig. 4.15b; [#k4], 763.43/138.62). Chloritoid is best recognized in thin sections with a petrographic microscope: light blue-green in plane polarized light (Fig. 4.15c) and high birefringence compared with chlorite in crossed polarized light (Fig. 4.15d). Chloritoid replaces staurolite from an older phase of metamorphism during the Alpine greenschist facies event (Wenk 1974).

Continue to Pass da la Duana (2694 m) through granitoid Suretta augengneiss (below the metasedimentary layers). From there descend on a good trail to Cadrin,

Fig. 4.14 **a** Large outcrops of Triassic marble with Karst morphology at Lägh da la Duana (hammer length is 50 cm). **b** "Edelweiss" (*Leontopodium alpinum*) associated with calcareous rocks. **c** View on Piz Duan from N. **d** View from summit of Piz Duan into Val Bondasca with Pizzo Badile (right) and Pizzo Cengalo (left). On the right side Monte Rosa in the far background

then Löbbia and Plän Vest (1821 m). Plän Vest is a big meadow on moraines. At the lower edge the trail passes again Triassic marbles on top of quartzites [#k5], this time of the lower Tambo nappe. There is an old lime kiln in the forest west of the trail (762.02/135.41).

After Plän Vest the trail descends fairly steeply through Tambo nappe gneisses towards Soglio. At 1550 m you pass through another alp, Tombal, also here on a wide moraine deposited by the valley glacier during the ice age. At the first house you should look towards the west, where under Cävi in the gorge Drögh Grand you can see a zone with white rocks ([#k6], 760.43/135.82, Fig. 4.15e, with detail in Fig. 4.15f). They also belong to the Tambo Triassic sediments observed

Fig. 4.15 a View southeast with Piz Duan and Piz Bernina in the background. Light rocks below Piz Duan are Triassic marble and quartzite. Below (dark grey) on the right side Suretta pelitic schists and gneisses. **b** Hand specimen of paragneiss with garnet (G) and chloritoid which is easily recognized in thin sections (**c, d**) (width 5 mm). **c** Plane polarized light: green-blue is chloritoid (C), white is muscovite (M). **d** Crossed polarized light: high birefringence of chloritoid compared with chlorite. **e** View from Tombal west towards Cävi with zone of marble and gypsum (white). **f** Same zone below Cävi for details

at Plän Vest. It is not only marble, but also gypsum (Fig. 3.1e). Gypsum is rare in metamorphic rocks because it transforms easily. It was originally formed in tropical lagoons and isolated salt pans, as happens today, for example, on the edge of the Dead Sea.

In the creek at the E-end of the Tombal meadow ([#k7], 762.52/135.14), where the moraine material is exposed, you find boulders of granite from Piz Grevasalvas and serpentinite from the Lunghin Pass in the moraine.

Another beautiful old lime kiln can be found at the west end of Tombal ([#k8], 762.01/135.12, Fig. 3.19c). We recommend taking a side-trail to descend that goes by the kiln. You will meet the marked trail again that continues to Soglio.

From Soglio you can use the Postbus to return to Casaccia (via Promontogno). In case you stay in Soglio, you could use the following day to visit the quarries of Tambo gneiss below Soglio ([#k9], 761.75/134.20) or Promontogno ([#k10], 763.09/134.37) (Vasin et al. 2017). Quarry managers usually give permission to visitors interested in geology. In the Soglio quarry show them the picture of former owner Dolfo Schumacher (Fig. 3.20a) and in Promontogno a picture of the quarry in this book (Fig. 3.20b). Hardly anywhere else can you find rocks with such perfect cleavage over large areas. These platy gneisses were originally granites that intruded in the Carboniferous period. During the Alpine thrusts they were deformed and recrystallized. Some of the granitic structures are preserved in outcrops along the Liro River north of Chiavenna (excursion 4.1.7, [#g5], Fig. 4.8g).

4.2.5 Roticcio-Val Furcela-Val da Cam-Piz Cam: Young Diabase Dikes, Stratigraphy of Suretta Nappe, Mn-Silicates (Map 3, waypoints #l1–#l10) (6–8 h)

A suitable parking space can be found at the NE-side of Roticcio ([#*l*1], 769.34/ 136.88). Then follow the trail uphill northwest through meadows and forest. At 1516 m you cross a road and take it because it has outcrops of albite-chlorite gneiss, so-called greenschist or prasinite of the Lizun unit belonging to the Platta nappe [#*l*2]. From Nambrun follow the trail on a moraine ridge to Alp Furcela (1942 m).

From here make a detour to a debris field under cliffs in the north ([#*l*3], 768.5/ 138.2). It is mainly gneiss, but locally you can find dark fine-grained dikes of diabase (Fig. 4.16a). These are young (~ 40 My) volcanic dikes, slightly older than Bergell granite. Outcrops are difficult to find e.g. above A. Furcela ([#*l*4], 768.5/ 138.3), south of Piz Lizun ([#*l*5], 769.3/139.3) and north of Piz Lizun ([#*l*6],

4.2 Longer Excursions (1–2 days)

769.2/139.5, Fig. 4.16b) for those who are interested in following up (Nievergelt and Dietrich 1977; Wenk 1980). Temperatures at depth were obviously high enough to form melts before Bergell granite intruded.

Now we take the path in Val Furcela to the pass Bocchetta da la Furcela ([#/7], 767.47/139.14, 2369 m). On the left is Piz Cam with white Triassic marble layers of the Suretta nappe and Jurassic calcareous schists of the Avers nappe (Fig. 4.16c). On the right side are greenschists (prasinite) which we observed earlier on the road. At the pass we also find serpentinite of the Platta nappe. We now use tracks and follow more or less the ridge towards the summit of Piz

Fig. 4.16 a Block with diabase dike in greenschist near Alp Furcela.(width 20 cm). b Vertical mafic dike in gneiss west of Piz Lizun above Maroz Dora. c View towards west with Piz Cam on the right side, mountains in the Ticino and behind Monte Rosa. The rocks on the right side are mainly Bündnerschiefer of the Avers nappe resting on marble (white) of the Suretta nappe. d View from Piz Cam east on Piz Lizun and Piz Lunghin with the Upper Engadine in the background

Cam. The calcareous and clayey rocks form soils for a rich flora. Beneath Piz Cam on the west side there is a zone with red manganese minerals (kutnahorite, rhodonite, rhodochrosite and piedmontite) [#*l*8]. These manganese minerals are an indication that the rocks originated in a deep ocean environment.

Piz Cam is also a beautiful panoramic mountain ([#*l*9], 767.12/137.71, 2633 m): There is a good view on Piz Lizun with the Engadine in the background (Fig. 4.14d). We descend to the west into the upper Val da Cam and cross a layer of quartzite and marble at Lägh da Cam ([#*l*10], 766.26/137.54) at the base of the Bündnerschiefer (Avers nappe). The white rocks were once limestones on top of sandstones and formed in the Triassic period about 240 million years ago. They are used as a guiding horizon in large parts of the Alps, displaying boundaries between nappes (e.g. Fig. 2.9a).

On the descent from Plan Lo to Sdarva and Bleis we reach a horizontal hiking trail (Via Panoramica, excursion 4.2.10), on which we return to Roticcio. At the stream crossing 768.97/136.87 we observe again the same marble zone of the Suretta nappe that we saw at Lägh da Cam.

4.2.6 Val Albigna, Pass da Casnil, Piz Casnil or Piz Bacun, Passo Cacciabella Sud, Pizzo Eravedar. Bergell Granite, Pegmatite and Aplite Dikes, Xenolith Inclusions, Orbicular Granite (Map 5, waypoints #m1–#m10) (2 days)

Day 1: First, follow the recommendations for excursion 4.1.4 from the Pranzaira cable car station to Capanna Albigna SAC [#d1–#d3].

From Capanna Albigna (Fig. 4.17a, [#d4], 770.68/133.37) we climb on the blue-white marked trail towards Pass da Casnil Sud, partially on moraines and some on outcrops. At first granite with orthoclase megacrysts dominates, but the higher we advance the more dark inclusions we observe (Fig. 4.17b) and also homogeneous granite without megacrysts. There are numerous outcrops with glacial polish, demonstrating that erosion has been minimal in the last 10,000 years.

At about 2450 m ([#m1], 771.05/133.51) the path passes through a zone of red soil of altered granite that was formed in fractures (Fig. 4.17c). A detailed analysis reveals that this red zone is mainly composed of the zeolite mineral laumontite (Fig. 3.2i).

A good place to study the different varieties of Bergell granite is a relatively flat plateau at 2560 m ([#m2], 771.37/133.71). There is still the normal granite

Fig. 4.17 Albigna. **a** Capanna Albigna (SAC). **b** On trail above to Pass Casnil large inclusion of tonalitic material with aplite dike in Bergell granite. **c** Along the trail at 2350 m red zone of alteration of granite to laumontite (width 40 cm). **d** Fine-grained granite with aplite vein crosscut by pegmatite (knife is 9 cm long). **e** At 2450 m orbicular structures similar to those observed below Monte del Forno (Fig. 4.9g). **f** View towards Piz Casnil (left) and Pass Casnil Nord with large moraines. **g** View from Piz Bacun towards Val Forno and Monte Disgrazia in the background

with megacrysts, but much of the granite is more homogeneous with few large feldspar crystals (Fig. 4.17d). There are many dark inclusions of hornblende-rich tonalitic rock (xenoliths) and occasionally even orbicular granites with spherical inclusions (Fig. 4.17e) which we described on the Forno excursion (4.2.1, Fig. 4.10g). Here in Val Albigna, we are obviously near the "roof" of granite and close to contact with amphibolite, as in the eastern Forno zone.

From here you can follow the main trail to the Pass da Casnil Sud and have a view of Val Forno. The marked trail continues into Val Forno on the east side. From the pass we recommend branching off on an unmarked trail at 2950 m to the north to get to Pass da Casnil Nord ([#m3], 772.75/133.78, 2971 m) without difficulty. You pass through large moraines (Fig. 4.17f) and outcrops with many variants of granite. From Pass da Casnil Nord it is easy to reach Piz Casnil 3187 m via the southwest side ([m4], 772.86/134.11).

Another option is to climb Piz Bacun instead of Piz Casnil. On the trail described above, at about 2550 m [#m2], you turn off towards north and then head up a side-valley with some small lakes to the NE. Continue to Furcela dal Riciöl (3039) ([#m5], 772.78/134.59). Descend ~50 m on the east side and then climb up a couloir to the north and from the crest descend slightly towards NW. Follow tracks towards Piz Bacun and climb it from the NE-crest ([#m6], 772.72/134.79). Piz Bacun is one of the highest mountains (3243 m) with a great view, including Monte Disgrazia (Fig. 4.17e).

For either variant return on the same path you came to the Capanna Albigna (SAC) [#d4].

Day 2: From the SAC hut take the renovated path that descends south towards the lake. Cross the stream on a bridge ([#m7], 770.72/132.96) and traverse at the bottom of cliffs under Punta da l'Albigna on the moraine and then descend to the glacier (Fig. 4.18a). At the northern end of the glacier we find large deposits of clay (Fig. 3.5e). Moraines, in contrast to river alluvium and talus debris, have a greater clay content and are therefore more fertile for plants.

Spend an hour studying the different blocks in the main moraine [#m8]. If you are lucky, you will find blue beryl crystals in pegmatites. Of interest is the large fraction of hornblende gneiss contained in the moraine. This rock resembles tonalite (Italian "Serizzo"), which occurs in Bagni del Màsino at the base of Bergell granite (Italian "Ghiandone") (Fig. 3.3c). With the retreat of the glacier a window of tonalite may become exposed ([#m9], 769.42/130.51), supporting the model that has Bergell granite resting on tonalite and not extending to great depth in Val Albigna (Fig. 2.8).

After crossing the glacier we climb on the blue-white marked trail up to Passo Cacciabella Sud (2896 m, [#m10], 768.81/131.79). During the ascent you

Fig. 4.18 a View on upper Val Albigna and Sciora group (right), indicating extent of old glacier (lighter colors). b View from Piz Cacciabella into Val Bondasca with glacier, Pizzi Gemelli and Pizzo Cengalo (right) (courtesy W. Hunkeler)

will again find outcrops of homogeneous granite (without feldspar megacrystals), which extends as a cylindrical structure throughout the whole massif, from Bondasca to Forno (maps 4 and 6). From the pass take a short hike north on the crest to Pizzo Eravedar (2933 m) and enjoy the view on the Sciora mountains, Pizzi Gemelli, Pizzo Cengalo, Pizzo Badile and Val Bondasca (Fig. 4.18b). The trail on the W-side leads into Val Bondasca and Capanna Sciora, SAC, has been temporarily closed after the 2017 Cengalo landslide. It is due to reopen in 2025.

On the descent on the E-side follow the blue-white marked trail north to the EWZ building at the lake and then down to the cable car station.

4.2.7 Lavinair Crusc-Piz Salacina: Calcareous-Silicates in Contact Zone (Map 1, waypoints #n1–#n7) (5–7 h)

What happens when granite comes into contact with marble? Granite is rich in Si and Al, marble in Ca and Mg and therefore there will be chemical reactions such as

$$SiO_2 \text{ (quartz)} + CaCO_3 \text{ (calcite)} \rightarrow CaSiO_3 \text{ (wollastonite)} + CO_2$$

In a similar way, other calcareous silicates form such as diopside $CaMg[Si_2O_6]$, scapolite $Ca_8[(Cl_2,SO_4,CO_3)_2/(Al_2Si_2O_8)_6]$, grossular garnet $Ca_3Al_2[SiO_4]_3$ and tremolite $Ca_2Mg_5[OH/Si_4O_{11}]_2$.

Good samples of calcareous silicates can be found in upper Lavinair Crusc where granite is in contact with Triassic calcite and dolomite marbles. Be prepared: This excursion is not along established trails, at least in the ascent (Fig. 4.19a), but mineralogically it is very interesting.

From the valley road, turn east south of Löbbia (770.89/138.06) and park at Ca d'Faret [#n1]. From here you hike east, first on a trail to the bottom of Lavinair Crusc (1700 m, [#n2], 771.87/138.12). Then it is best to continue in river gravel where you get an overview of metamorphic contact rocks. At ~2000 m you get good samples of marble with red grossular garnets ([#n3], Fig. 4.19b) and at ~2400 m there are outcrops of amphibolite crossed by pegmatite. We climb up to the ridge (~ 2500 m, [#n4], 773.12/138.59) and from there make a detour south along the ridge towards the base of Piz Murtaira, first through amphibolites

Fig. 4.19 **a** View on Piz Salacina (center) and the Lavinair Crusc canyon with large debris slopes. **b** Scrambling up the canyon we find many metamorphic reactions in the contact zone such as grossular garnet crystallizing in marble (width 30 cm). **c** On top of the ridge towards Cima di Murtaira contact of schist and Bergell granite with large feldspars indenting the schist (width 35 cm). **d** Chloritized amphibolite in Motta Salacina (hammer length 50 cm) (**b** courtesy of W. Hunkeler)

4.2 Longer Excursions (1-2 days)   107

and paragneiss, then to the granite contact with large feldspars from the granite pushing into micaceous layers of the gneiss ([#n5], 773.17/137.91); Fig. 4.19c).

We return to the pass and follow a trail to the summit of Piz Salacina (2600 m, [#n6], 773.04/138.71), mainly through amphibolite. Piz Salacina gives a panoramic overview of the Val Forno with Tertiary Bergell granite (SE), the stack of Pennine nappes (W, Piz Duan; "Europe"), Austroalpine nappes (E, Piz da la Margna, Piz Fedoz; "Africa"), and in between rocks such as amphibolite which indicate an oceanic origin (Tethys, "Mediterranean").

There are two options for descent as indicated on the map. Either we follow the N-crest for 200 m and then descend on a trail to Starlögia or we return on the S-crest to the saddle and from there pass on meadows to Starlögia. There are cattle trails to Pass dal Caval with amphibolite ([#a9], excursion 4.1.1). Amphibolite is often converted into chlorite epidote schist (Fig. 4.19d). Over local moraines without Bergell granite we come to Motta Salacina ([#n7], 772.67/139.59, Fig. 4.1a) and on a good trail descend to Löbbia and from there on a trail through the forest to Ca d'Faret.

4.2.8 Bondo-Ciresc-Lera d'Sura-Denc dal Luf: Al$_2$SiO$_5$ Triple Point, Gruf Migmatites, Ultramafics, Moraines (Map 7, waypoints #o1–#o8) (5–8 h)

This fieldtrip is on the S-side of Val Bregaglia, mainly forested with red and white firs and larches (Fig. 4.20a). Moraines, rockfalls and talus form the morphology. Rocks are granitic gneisses, metasedimentary para-gneisses with interesting mineral assemblages (from the Tambo nappe), a zone of ultramafic schists (of the Chiavenna zone) and at the top migmatites (Gruf unit).

In Bondo park your car above the village ([#o1], 762.53/133.44) and then take the forest road to Ciresc. It passes through a steep talus slope with forest that stabilizes erosion. Where too many trees were lumbered a dangerous slide occurred in 1995 (Fig. 4.20b, [#o2]). In Ciresc (1288 m) you can find glacial moraines, ideal for agriculture, but without Bergell granite. Thus they were from local glaciation.

From Ciresc proceed along the horizontal forest road to Cugian. On the way there are geologically interesting locations, but many outcrops are overgrown and covered with lichens. Rocks are biotite-muscovite mica schist with garnet and aluminosilicates. From coordinates 762.0/132.6 to 762.4/132.8 [#o3] there are still a few outcrops of these pelitic schists, but a hammer is required to recover fresh rocks. On closer examination, it was observed that in the same

Fig. 4.20 **a** View from Soglio on the S-slope of Val Bregaglia with Piz Badile left (E) and Cima di Codera on the right (W). Forested slope indicating complex geomorphology. **b** Road Bondo-Ciresc in 1995 showing a rockslide after lumbering. **c** View on Denc dal Luf from SW. **d** Denc dal Luf above Alpe Lera d'Sura. **e** Blocks of migmatite (grey-white) and ultramafic gneiss (brown). **f** Moraine surface in Ciresc with view into Val Bregaglia. **g** Old hole in ultramafic rock in Ciresc (diameter 20 cm) (**a, d, e** courtesy of W. Hunkeler)

and paragneiss, then to the granite contact with large feldspars from the granite pushing into micaceous layers of the gneiss ([#n5], 773.17/137.91); Fig. 4.19c).

We return to the pass and follow a trail to the summit of Piz Salacina (2600 m, [#n6], 773.04/138.71), mainly through amphibolite. Piz Salacina gives a panoramic overview of the Val Forno with Tertiary Bergell granite (SE), the stack of Pennine nappes (W, Piz Duan; "Europe"), Austroalpine nappes (E, Piz da la Margna, Piz Fedoz; "Africa"), and in between rocks such as amphibolite which indicate an oceanic origin (Tethys, "Mediterranean").

There are two options for descent as indicated on the map. Either we follow the N-crest for 200 m and then descend on a trail to Starlögia or we return on the S-crest to the saddle and from there pass on meadows to Starlögia. There are cattle trails to Pass dal Caval with amphibolite ([#a9], excursion 4.1.1). Amphibolite is often converted into chlorite epidote schist (Fig. 4.19d). Over local moraines without Bergell granite we come to Motta Salacina ([#n7], 772.67/139.59, Fig. 4.1a) and on a good trail descend to Löbbia and from there on a trail through the forest to Ca d'Faret.

4.2.8 Bondo-Ciresc-Lera d'Sura-Denc dal Luf: Al_2SiO_5 Triple Point, Gruf Migmatites, Ultramafics, Moraines (Map 7, waypoints #o1–#o8) (5–8 h)

This fieldtrip is on the S-side of Val Bregaglia, mainly forested with red and white firs and larches (Fig. 4.20a). Moraines, rockfalls and talus form the morphology. Rocks are granitic gneisses, metasedimentary para-gneisses with interesting mineral assemblages (from the Tambo nappe), a zone of ultramafic schists (of the Chiavenna zone) and at the top migmatites (Gruf unit).

In Bondo park your car above the village ([#o1], 762.53/133.44) and then take the forest road to Ciresc. It passes through a steep talus slope with forest that stabilizes erosion. Where too many trees were lumbered a dangerous slide occurred in 1995 (Fig. 4.20b, [#o2]). In Ciresc (1288 m) you can find glacial moraines, ideal for agriculture, but without Bergell granite. Thus they were from local glaciation.

From Ciresc proceed along the horizontal forest road to Cugian. On the way there are geologically interesting locations, but many outcrops are overgrown and covered with lichens. Rocks are biotite-muscovite mica schist with garnet and aluminosilicates. From coordinates 762.0/132.6 to 762.4/132.8 [#o3] there are still a few outcrops of these pelitic schists, but a hammer is required to recover fresh rocks. On closer examination, it was observed that in the same

Fig. 4.20 **a** View from Soglio on the S-slope of Val Bregaglia with Piz Badile left (E) and Cima di Codera on the right (W). Forested slope indicating complex geomorphology. **b** Road Bondo-Ciresc in 1995 showing a rockslide after lumbering. **c** View on Denc dal Luf from SW. **d** Denc dal Luf above Alpe Lera d'Sura. **e** Blocks of migmatite (grey-white) and ultramafic gneiss (brown). **f** Moraine surface in Ciresc with view into Val Bregaglia. **g** Old hole in ultramafic rock in Ciresc (diameter 20 cm) (**a, d, e** courtesy of W. Hunkeler)

rock three minerals of the same composition Al_2SiO_5 occur together: andalusite (pink), kyanite (blue) and sillimanite (white needles) (Fig. 3.11b, see Sect. 3.3 and Wenk et al. 1974). The minerals define a "triple point" for pressure–temperature conditions during Alpine metamorphism (T = 520 °C, P = 400 MPa; Fig. 3.11a).

In Cugian take the trail to Luvartigh and Lera d'Sura. On the way there are good views into Val Bondasca, with the Sciora Group, Pizzo Cengalo and Pizzo Badile. Particularly impressive are views of the debris from the 2017 landslide in upper Val Bondasca ([#o4], 763.30/132.03, Fig. 3.17a, b). On the way up you pass through outcrops of micaschists of the Tambo nappe and occasionally layers of amphibolite.

From the alp Lera d'Sura ([#o5], 762.96/131.93, 1894 m) walk more or less horizontally on a cattle trail into the basin on the southeast side and remain below the large brown block field (763.08/131.71, 1890m).

From here we climb through the grassy slope up towards the left side of Denc dal Luf ("Wolf's Tooth", 2171 m), a prominent cliff visible from afar, all the way from Chiavenna ([#o6], 762.69/131.39, Fig. 4.20c, d), to ~2000 m. We explore the large talus field with big blocks of grey migmatites on the left (south) and brown ultramafic rocks on the right (north). The migmatites are rocks of granitic composition that have been mobilized but not completely melted during various episodes (e.g. Galli et al. 2013). They contain mobilized Tertiary pegmatite and aplite veins. The ultramafic rocks are part of the lavez zone with talc-olivine-chlorite-amphibole-diopside-serpentine schists which extends from Chiavenna into Val Bondasca (see Sect. 3.6.1 and Schmutz 1976). The brown color is a surface alteration. The fresh samples are dark green. Traverse to the north towards the base of Denc dal Luf.

For mountaineers Denc dal Luf is challenging but not very difficult: You climb up with a rope on the north side and then rappel from the summit on the south side.

Now we return to Lera d'Sura and then on the marked red-white trail to the west. Serpentinite and ultramafic rocks are often mechanically unstable and associated with landslides (e.g. Piuro/Plurs). The large debris field on the northwest side of Denc dal Luf is one such rockfall named "Ganda Rossa" because of the reddish color of the weathered olivine-talc rock ([#o7], 761.94/131.76). The morphology with talus slopes and more stable rocks is expressed in the forest with light green larch trees on unstable surfaces and dark green fir trees on more stable locations as can be seen in the view from Soglio (Fig. 4.20a).

We follow the path to Vec, Cänt and Lizöl to Ciresc, with many meadows on moraines ([#o8], 7761.30/132.38), Fig. 4.20f). Near upper Ciresc is a handcrafted hole in ultramafic schists which may have been used for grinding in ancient times (Fig. 4.18g). From here we return on the forest road to Bondo.

4.2.9 Bondo-Cugian-Trubinasca-Cap. Sasc Fura (Val Bondasca). Contact Zone of Bergell Granite, Ultramafic Xenoliths, Deformed Granite, View of the 2017 Landslide (Map 7, waypoints #p1–#p5) (7–9 h)

One of the geologically most fascinating places in the Bergell is Val Bondasca where you can experience a three-dimensional impression of the geological structures (Fig. 2.8) and explore the different types of Bergell granite: megacrystalline granite, fine-grained homogeneous granite without alkali feldspar megacrysts, tonalite, xenolith swarms, deformation of granite, etc. Unfortunately, the valley is still largely closed due to the 2017 Cengalo rockfall but a new trail from Promontogno over Prä and Naravedar to the Capanna Sciora is under construction. The Alp Laret with beautiful old huts and the Sciora mountains in the background are fond memories (Fig. 4.21a). Val Bondasca has a famous reputation for technical climbing such as the Badile northeast face and N-crest (Fig. 4.21b as well as the book cover) and the peaks of the Sciora group, especially the spike Ago di Sciora (Fig. 4.21c). Here we describe an excursion on the southside of Val Bondasca.

Park above Bondo [#o1] then on foot advance on the forest road. At an intersection choose Val Bondasca (left), not Ciresc (right). We traverse a tunnel with Tambo augengneiss that was once a megacrystic granite ([#p1], 762.95/133.37). The granitic variety is partially preserved in the river Liro above Chiavenna as described in excursion 4.1.7 [#g5, Fig. 4.8g].

At the new bridge (1023 m, [#p2], 763.45/133.16), built after the 2017 landslide destroyed the old bridge, look at the river in the gorge and polished rock surfaces. From here we take the trail to Cugian on a moraine shoulder, then we continue to Luvartigh. From Luvartigh (1553 m) we follow the new blue and white marked trail through Predacia to Valun da la Trubinasca. This was previously an old hunter's path where the last wild bear in the Bergell was shot in 1867 and can still be viewed in the Stampa valley museum. The path goes through rocks of Gruf migmatite, a gneiss which has been partially mobilized, often with folds ([#p3], Fig. 4.21d) and crosscut by pegmatites.

In upper Trubinasca valley we climb to 2250 m and reach the contact zone of the Bergell tonalite, a hornblende-rich variety of Bergell intrusives This young igneous rock lies above the metamorphic rocks, and surrounds the megacrystic variety like a blanket (Fig. 2.8). In the contact zone under the tonalite one finds ultramafic rocks of the Chiavenna olivinite-talc schist-serpentinite zone which have been transformed with impressive zoning and represent a beautiful example of metasomatism ([#p4], 764.52/130.39). Figure 4.21e shows Friedrich Drescher-Kaden in 1969 at the outcrop with transformed breccias. His research

4.2 Longer Excursions (1–2 days)

◀**Fig. 4.21 a** View east from Sasc Furä in Val Bondasca on the Sciora group before the 2017 landslide. Left is Sciora Dafora, center Ago di Sciora and right Sciora Dadent. **b** A view on Pizzo Badile's NE-face and N-crest in Val Bondasca. They are some of the most adored climbs in the Alps. **c** View from the bottom of the northeast face of Pizzo Badile on the Sciora group Dafora, Ago and Dadent. **d** Typical migmatites observed on the trail from Luvartigh into upper Val Trubinasca (hammer length 50 cm). **e** Prof. Drescher-Kaden exploring metasomatic breccias in Trubinasca glacier (1969). **f** Partially transformed ultramafic breccias in Val Trubinasca (width 30 cm). **g** Bergell granite on the slope above Capanna Sasc Furä (SAC). **h** Swarm of xenoliths in Bergell granite at the base of Bondasca glacier. **i** Strongly deformed foliated Bergell granite with stretched xenoliths. Base of Pizzi Gemelli

was largely dedicated to the investigations of the effect of hydrothermal alteration (e.g. Drescher-Kaden 1940, 1961) and as mentioned earlier he was one of the pioneers investigating Bergell granite (Drescher-Kaden and Storz 1926). The center of the spherical bodies is predominantly olivine, which has been converted into pyroxenes, amphiboles and talc under the influence of silica-rich solutions (Fig. 4.21f).

From here the trail descends to Capanna Sasc Furä SAC, the starting point for climbing expeditions on Pizzo Badile. Slightly above the hut we find the contact migmatite-tonalite and higher tonalite-megacrystalline granite, with a good trail to 2265 m ([p5], 765.43/130.79, Fig. 4.21g). Above it climbing begins and on this trip we turn around!

The path traversing to the Capanna Sciora is closed due to the 2017 landslide. There are magnificent outcrops at the base of Pizzo Cengalo and Pizzi Gemelli with large swarms of xenoliths (Fig. 4.21h) suggesting that we are near the base of the granite body. These structures are very different from granite structures in Val Forno and Val Albigna (e.g. Figs. 4.5d and 4.10f), and underwent strong plastic deformation imposing large strains. Megacrysts are aligned like in an augengneiss and xenoliths are stretched out (Fig. 4.21i).

The trail from the Sasc Furä into Val Bondasca is still officially closed. We therefore have to return to Bondo the same way we came.

4.2.10 Overview: Via Panoramica from Casaccia to Soglio (Map 3, waypoints #q1–#q6) (4–6 h)

We assume that you have become familiar with the wide variety of Bergell rocks on some previous excursions with hammer and hand lens. This hike serves as a review with views on places you have already visited, and it offers a good

4.2 Longer Excursions (1–2 days)

opportunity to digest the observations. Also, as you will see, there are not many rock outcrops along the Via Panoramica because a lot is covered by moraines and rockfalls. You may wish to review the geological profile of the EWZ tunnel from Löbbia to Castasegna that goes more or less parallel to the trail, underneath the surface (Fig. 3.23).

There is a parking lot just outside Casaccia (#c1). Let's look first to the northeast with Piz Lunghin (Fig. 3.16a). These rocks belong to the Austroalpine Margna nappe. You can see a small landslide under Piz dal Sasc from 1970. In 1673 there was a much larger one that covered the whole village of Casaccia and the ground floor of houses became cellars (Fig. 3.16b). If you have not done it before (excursion 4.1.3) take a brief walk through the old village. For example where the main valley road narrows there is a building on the E-side where the old entrance door along the highway is now the entrance to the cellar. It reminds us that rockfalls are among the most dangerous geological events in this valley with steep slopes.

We start on the Via Panoramica towards the south (Fig. 4.22a) and pass through a flat meadow between Casaccia and Löbbia. In the background is Val Albigna with the EWZ hydroelectric power dam. This large alluvial plane was created by the river Orlegna. As you walk along you see a gravel pit on the left where gravel and sand are exploited (Fig. 3.20d).

We cross the Maira and there is a small quarry of greenschist (prasinite) that was used for construction ([#q1], 770.84/139.46). These greenschists, where chlorite produces the green color, belong to the Lizun unit of the Platta nappe and they often contain small crystals of pyrite (FeS_2). We mentioned in Sect. 3.6.1 that in some places chalcopyrite ($CuFeS_2$) was found and exploited for copper production, e.g. in Motta Farun above Casaccia (Fig. 3.21c, d).

After a kilometer we come to a small lake and across it are the headquarters of the EWZ power plant of Löbbia which convert the energy of the water from Albigna and Forno into electricity ([#q2], 770.71/138.37, Fig. 3.22c). At times of low demand, water is also pumped up from here into Lake Albigna for storage. From Löbbia there is a fairly flat tunnel to Castasegna where electricity is generated again. This tunnel is approximately parallel to the Via Panoramica. But while the Via Panoramica has very few rock outcrops and is largely on talus and moraines, the tunnels are 100% in rocks and thus provide an excellent geological background for this hike (Fig. 3.23). There are some outcrops of schists of the Suretta nappe at the Löbbia dam.

Via Panoramica continues, mainly on moraines with Bergell granite over Foppa, Barga and Pisnana to Roticcio. These moraines were at the base of the Pleistocene Bergell Glacier, which transported Bergell granite from Val Forno

Fig. 4.22 Via Panoramica from Casaccia to Soglio. **a** Large alluvial plane between Casaccia and Löbbia. **b** Outcrop of marble at river crossing south of Roticcio. **c** Old moraine at Durbegia with view towards Pizzo Badile (left) and Cima di Codera (right). **d, e** At waterfall outcrops of Suretta augengneiss (knife is 9 cm long). **f** Closer to Soglio outcrops of platy Tambogneiss which is used for steps, with view into Val Bondasca. **g** Evidence of glacial polish on the trail. **h** A last look east into Val Bregaglia, a really panoramic view. **i** Descent into Soglio along the old trail

across Chiavenna to the end of Lake Como. From Roticcio we have a good view towards Albigna with challenging climbing peaks, including Fiamma, Spazzacaldera, Cacciabella, Cima dal Largh and Piz Cantun, all in Bergell granite (excursion 4.2.6).

4.2 Longer Excursions (1–2 days)

Soon after Roticcio we cross a stream with outcrops of marble and a bit later quartzite of the Suretta nappe which were once limestones and sandstones of the Triassic period ([#q3], 768.97/136.87, Fig. 4.22b). Check with your hammer: it scratches marble but not quartzite. A bit further the path crosses Suretta augengneiss underneath the marble and originally granite. Remember this is the same marble and gneiss which we observed in Val da la Duana (excursion 4.2.4) and below Piz Cam (excursion 4.2.5). The path continues mainly through the forest with neither much visible geology nor views until we reach the meadow Durbegia above Vicosoprano ([#q4], 766.17/135.85). This offers views east, south and west (Fig. 4.22c). Explore large blocks, and boulders around them. The large blocks are gneiss with sharp faces indicating a rockfall origin but small rounded boulders were deposited by the moraine. There are even a few small boulders of Bergell granite.

We continue in the forest to the west, first on a dirt road, then on a trail. The path crosses two creeks with waterfalls ([#q5], 764.68/136.01, Fig. 4.22d). These outcrops are also augengneiss from the Suretta nappe (Fig. 4.22e). At Guäld above Pravis we pass a huge rockfall that covered the slope above Muntac. After Parlongh, with a view into Val Bondasca, there are again outcrops, some with glacial polish (Fig. 4.22f, g). We are now in the Tambo nappe and observe biotite micaschists, some amphibolites and below platy gneiss, also used for the construction of the trail ([#q6], 762.90/134.83, Fig. 4.22f).

Before we descend to Soglio you can once again enjoy the view south into Val Bondasca with the Sciora Group (Fig. 4.22f) and east into Val Bregaglia (Fig. 4.22h). Above Soglio you have a view of the old trail, the village and the Italian mountains in the background (Fig. 4.22i). Today, the majority of the hillside is forest, including most of Via Panoramica. This has changed a lot in 100 years. At that time there were many more meadows and pastures that were cultivated.

From Soglio you can take the postbus via Promontogno back to Casaccia or investigate the gneiss quarries at Soglio and Promontogno (Fig. 3.20a, b).

4.2.11 Bondo-Ciresc-Tegiola-Val Codera-Novate Mezzola: Gruf Migmatite, Mylonites, High Temperature Mineral Assemblages, Novate Granite (Map 8, waypoints #r1–#r7) (2 days)

This arduous two-day tour with more than 2000 m difference in elevation offers insight into some of the most extraordinary rocks of the Alps and at the same time into a wild mountain landscape that already fascinated Studer (1851). We start again in Bondo [#o1] and follow the road to Ciresc (excursion 4.2.8).

From Ciresc a trail leads over Cänt and Vec to Alpe Tegiola. At 1580 m, at an idyllic spring, you cross an ultramafic zone extending from Chiavenna with talc-olivine schists ([#r1], 761.42/131.56). After Alpe Tegiola you are in Gruf migmatites with granitic composition. There is isoclinal folding, banding with layers of slightly different composition and granitic veins suggesting partial melting (Fig. 3.3d). Below Bocchetta della Tegiola one finds numerous mylonite and ultramylonite layers ([#r2], 7762.59/130.10, Fig. 4.23a). These dark, flint-like, fine-grained rocks display an intense deformation in thin sections (Fig. 4.23b). The extreme deformation may be related to the emplacement of the Bergell granite. A narrow branch of Bergell granite extends deep into Gruf migmatites westwards along a fracture and ends along the ridge towards Cima di Codera.

Descending from the pass (2490 m) you enter Val Codera. A view from Pizzo di Prata east displays a wild valley surrounded by mountains with cliffs and steep slopes (Fig. 4.21c). A view towards Pizzo Ligoncio reveals the tectonic structure already observed by Studer (1851) on a hike from Val Codera to Bagni del Masino: "Dark schists at the bottom, above hornblende rocks, fine-grained granite in the center and porphyritic granite on top ..." which can be translated into: metamorphic schists and migmatites at the bottom, superposed by tonalite, fine-grained granite, and megacrystalline Bergell granite on top (Fig. 4.23d).

It is possible to spend a night in Bivacco Pedroni Del Pra (formerly Bivacco Vaninetti) at the foot of Pizzo Trubinasca ([#r3], 7763.86/129.37, 2577 m) or descend directly into the valley. The detour is worthwhile. Under and above the Bivacco there are excellent outcrops of migmatite, aplite and pegmatite dikes (partly with garnet and beryl), as well as ultramafics with epidote and large blue cordierite crystals in biotite schists (Fig. 4.23e) in the contact zone with Bergell granite. On the descent into Val Codera you will find in the creek bed about 200 m south of Alpe Sivigia ([#r4], 762.89/128.96) the same ultra-mafic breccia as in Trubinasca (excursion 4.2.9, c.f. Fig. 4.21f).

From the side valleys Valle del Conco and Val Piana blocks of aluminum-rich schists with sillimanite, garnet, hypersthene, cordierite, corundum and rarely sapphirine were deposited and can be observed along the trail. This mineral association is typical of the deepest crust and formed at high pressure and high temperature. These rocks of the so-called granulite facies (Fig. 2.7) are most accessible at the classical Cornelius (1916) site ([#r5], Fig. 4.23f, 760.9/125.3) in debris under a rock face. With this mineral diversity Val Codera has become a treasure trove for mineralogists (e.g. Ghizzoni and Mazzoleni 2005).

4.2 Longer Excursions (1–2 days) 117

◀**Fig. 4.23** Passo Tegiola, Val Codera, Novate. **a** A subvertical dike-like structure of ultramylonite north of Bocchetta della Tegiola. **b** Microstructure with very fine-grained recrystallized quartz flowing around a feldspar crystal. Cross-polarized light microscopy (width is 5 mm). **c** View from Pizzo di Prata east into Val Codera. **d** S-side of Val Codera with Pizzo Ligoncio and Cime di Gaiazzo. Migmatite in the valley superposed by tonalite and Bergell granite, an observation Studer made in 1851. **e, f** High temperature metamorphic assemblages. **e** Blue cordierite and garnet near Bivacco Pedroni dal Pra (width 10 cm). **f** Sapphirine-garnet-cordierite fels near Bresciadega (width 15 cm). **g** Zircon in gneiss growing from 530 to 33 My (courtesy of Galli et al. 2012). **h** Granite quarries around Novate-Mezzola, 1956. **i** River pebbles in Novate give an overview of the complex geological history of Val Codera with granite, migmatites, tonalites and metamorphic rocks

The migmatites of Val Codera were studied by Galli et al. (2012, 2013) and they confirmed that they correspond to high granulite facies conditions (> 920 °C, 9 kbar), and crystallized over a long period beginning at ~530 My in the deepest crust and continuing to 33 My during the intrusion of Bergell granite, as documented by zoning of zircon and corresponding U/Pb ages determined with a microfocus ion microprobe (Fig. 4.23g). This zircon started to grow long before the "old" Alpine granites intruded, such as Aare, Gotthard and Bernina (~ 300 My). A particularly interesting rock is charnockite with quartz, plagioclase, alkali feldspar, garnet, orthopyroxene and sillimanite.

Capanna Luigi Brasca CAI and Rifugio Tartaglione Crispo also offer night accommodation and meals as an alternative to the more primitive Bivacco Pedroni-Del Pra.

On the long descent to Novate-Mezzola, one observes more and more microgranitic outcrops that are Novate granite (Fig. 3.3b). Novate granite (also known as San Fedelino granite) is slightly younger than the Bergell granite (20–25 My, e.g. Liati et al. 2000), homogeneous and provides excellent construction material for buildings. There were large quarries in the 1950s (Fig. 4.23h). Today granite extraction is much more limited and is best studied in the quarries east of the Codera river (road towards San Giorgio) ([#r6], 755.96/120.56). A relaxing day of rest can be spent in the creek bed near Novate-Mezzolpiano, where you can swim in a geological garden amidst polished blocks and review the geologic diversity ([#r7], 756.0/120.8, Fig. 4.23i).

4.2.12 Valmalenco: Serpentinite Near Chiesa, Contact Rocks with Tonalite in Val Sissone (Map 9, waypoints #s1–#s7) (6–8 h)

A day in Valmalenco gives a different perspective of the Bergell Alps geology: serpentinite. It was originally peridotite, mainly composed of olivine and pyroxene, and part of the upper mantle of the Earth. Later it was uplifted into the Mediterranean oceanic crust and transformed to serpentinite through reactions with hydrous solutions. Here serpentinite is composed of the high temperature polymorph antigorite, while in the north such as Piz Lunghin it is lizardite. In Valmalenco some olivine is still present. Monte Disgrazia, the highest mountain in the Bergell Alps (4678 m), consists of serpentinite, and serpentinite is intensively mined in Valmalenco near Chiesa.

Val Sissone, in the upper part of Valmalenco, displays a unique contact zone between marble, amphibolite, paragneiss with Tertiary Bergell tonalite. We recommend the study by Trommsdorff et al. (2005) for more details.

We start in Sondrio. On the west side of the city, in Campoledro, the SP15 road branches off and climbs up the mountain. The Insubric Line is immediately traversed (coord. 787.11/116.51). Then, north of Mossini one crosses a small lens of Tertiary tonalite (Triangia tonalite) ([#s1, not on map], 787.33/118.03). Follow the main road to Torre Santa Maria (coord. 786.20/123.39) and from there take the road "Via Roma" on the west-side of the river to Chiesa in Valmalenco. Continue on the road "Via Bernina" to the north. Along the road there are several quarries of serpentinite (e.g. Fig. 4.24a) and in Chiesa there is a large center for serpentine processing (Nuova Serpentino d'Italia) where a stop is recommended ([#s2, not on map], 785.88/127.89). It gives an insight into the different types of Malenco serpentinite, from tombstones to roof slabs, and allows you to collect samples. A large quarry is just behind the factory. Malenco sepentinite has long been the focus of many geological studies. It constitutes transformed peridotite from the upper mantle and is part of the Platta/Malenco zone between Austroalpine and Pennine nappes, associated with the Tethys ocean (Fig. 2.4).

From Chiesa follow the road to Chiareggio. At Chiareggio we recommend a visit to the Parco Geologico (Montrasio et al. 2005, https://www.sondrioevalmalenco.it/en/parco-geologico-di-chiareggio) on two hectares of land with many rocks and explanatory plaques ([#s3], 781.6/132.4). Plan an hour and a half for this visit.

Then park your car in Pian del Lupo near Hotel Cembro ([#s4], 780.09/131.83). We follow the path to Forbesina and Rifugio Tartaglione-Crispo which is also a place to stay. From here we take a 4-h hike into Val Sissone.

Fig. 4.24 a Serpentinite quarry Sasso Corvi-Chiesa, Valmalenco. b View on Monte Disgrazia and Val Sissone. c Contact zone of tonalite with calcareous and pelitic rocks at the foot of the Disgrazia glacier in Val Sissone. d Contact zone between amphibolite and pegmatites. e Details of contact between amphibolite (right) and tonalite (left) cut by pegmatites. f Complex contact structures at Rifugio del Grande Camerini

First take the path along the stream and cross the river to find on the south-side an outcrop of talc-olivine-tremolite fels, similar to those around Chiavenna, and used for production of lavez pots ([#s5], 779.15/130.92). Other Lavez mines in Valmalenco are on the east-side near Franscia/Valbrutta where an old workshop was retrieved which is now on display in the museum Ciäsa Granda in Stampa (Fig. 3.18b).

Next follow the trail west into Val Sissone. Here there is no Bergell granite but instead tonalite (in Italian often referred to as "serizzo"), the mafic variety with hornblende and only little alkali feldspar (Fig. 3.3c) which we observed in Val Bondasca and Val Codera. In the stream bed there are blocks of pelitic schists and marble and at 2050 m is an outcrop below the glacier to observe the contact of these metasedimentary rocks with tonalite ([#s6], 777.91/129.32, Fig. 4.24c). Marble contains wollastonite ($CaSiO_3$), tremolite ($Ca_2Mg_5[(Si_8O_{22})(OH)_2]$) and grossular garnet ($Ca_3Al_2[SiO_4]_3$). Mica schist has the rare mineral mullite and sillimanite (Al_2SiO_5) (Wenk 1983). Val Sissone is one of the best places to study and enjoy contact metamorphism.

After this stop, ascend on the trail northwards, traversing the tonalite-amphibolite contact zone, often with coarse hornblendites (Fig. 4.24d, e; also Fig. 3.3c with deformed inclusions of hornblendite in tonalite). At Rifugio Del Grande Camerini, CAI ([#s7], 777.41/131.19) we observe excellent contacts of tonalite and amphibolite, crosscut by pegmatites (Fig. 4.24f). From the hut admire the view of Monte Disgrazia (Fig. 4.24b). Despite its ominous name it is one of the most beautiful mountains in the region. Originally it was called by shepherds Pizzo Bello.

Now we descend on a path to Alpe Vazzeda, through amphibolite, rarely with pillow structures and chalcopyrite mineralization. From there take the marked trail to Pian del Lupo.

On the way back stop in Sondrio and visit the Museo dei Minerali, particularly the Grazioli Mineral Collection.

References

Bernoulli D, Weissert H (1985) Sedimentary fabrics in Alpine ophicalcites, South-Pennine Arosa zone, Switzerland. Geology 13:755–758

Cornelius HP (1916) Ein alpines Vorkommen von Sapphirin. Zbl Mineral 11:265–269

Cornelius HP (1950) Geologie der Err-Julier-Gruppe. II. Teil: Der Gebirgsbau. Beitr Geol Karte Schweiz [N.F.] 70(2)

Di Capua A, Vezzoli G, Cavallo A, Gropelli G (2015) Clastic sedimentation in the Late Oligocene Southalpine Foredeep: from tectonically controlled melting to tectonically driven erosion. Geol J 51(3):338–353

Drescher-Kaden FK, Storz M (1926) Ergebnisse petrographisch-tektonischer Untersuchungen im Bergeller Granit. N Jb Mineral Geol Paläont Beilageband A 54:284–291

Drescher-Kaden FK (1940) Beiträge zur Kenntnis der Migmatit- und Assimilationsbildungen sowie synantetischen Reaktionsformen. I. Über Schollenassimilation und Kristallisationsverlauf im Bergeller Granit. Chem Erde 12:304–417

Drescher-Kaden FK (1961) Olivin-Metasomatose in Carbonatgesteinen aus der Umrandung des Bergeller Granitmassivs (Oberengadin). Naturwissenschaften 48:300

Ferrario A, Montrasio A (1976) Manganese ore deposit of Monte del Forno. Its stratigraphic and structural implications. Schweiz Mineral Petrogr Mitt 56:377–385

Galli A, Le Bayon B, Schmidt MW, Burg J-P, Reusser E, Sergeev SA, Larionov A (2012) U–Pb zircon dating of the Gruf complex: disclosing the late Variscan granulitic lower crust of Europe stranded in the Central Alps. Contrib Mineral Petrol 163:353–378

Galli A, Le Bayon B, Schmidt MW, Burg J-P, Reusser E (2013) Tectonometamorphic history of the Gruf complex (Central Alps): exhumation of a granulite-migmatite complex with the Bergell pluton. Swiss J Geosci 106:33–62

Gansser A, Gyr T (1964) Über Xenolithschwärme aus dem Bergeller Massiv und Probleme der Intrusion. Eclogae Geol Helv 57:577–598

Ghizzoni S, Mazzoleni G (2005) Itinerari mineralogici in Val Codera. Geologia Insubrica, 316 pp

Handy MR, Herwegh M, Kamber BS, Tietz R, Villa IM (1996) Geochronologic, petrologic and kinematic constraints on the evolution of the Err-Platta boundary, part of a fossil continent-ocean suture in the Alps (eastern Switzerland). Schweiz Mineral Petrogr Mitt 76:453–474

Liati A, Gebauer D, Fanning M (2000) U–Pb SHRIMP dating of zircon from the Novate granite (Bergell, Central Alps): evidence for Oligocene-Miocene magmatism, Jurassic/Cretaceous continental rifting and opening of the Valais trough. Schweiz Mineral Petrogr Mitt 80:305–316

Maurizio R (1972) Indagini su vecchie cave e miniere in Bregaglia. Quad Grigioni Ital 41:1–71

Meier R, Alig P (2006) Alpenführer Bündneralpen Südliches Bergell. Klettern, Hochtouren und Alpenwanderungen. SAC Verlag, 656 pp

Montrasio A, Sciesa E (1988) Carta Geologica della Valle Spluga ed Aree Aidiacenti, 1:50,000. Consiglio Nazionale delle Ricerche, Univ. degli Studi di Milano

Montrasio A, Trommsdorff V, Hermann J, Müntener O, Spillmann P (2005) Carta Geologica della Val Malenco, 1:25,000. Consiglio Nazionale delle Ricerche, Univ. degli Studi di Milano

Nievergelt P, Dietrich V (1977) Die andesitisch-basaltischen Gänge des Piz Lizun (Bergell). Schweiz Mineral Petrogr Mitt 57:267–280

Nigg P (2004) Bergell Gebirgsführer für Wanderer, Bergsteiger und Kletterer. Bergverlag Rother, 356 pp

Peretti A, Köppel V (1986) Geochemical and lead isotope evidence for a mid-ocean ridge type mineralization within a polymetamorphic ophiolite complex (Monte del Forno, North Italy/Switzerland). Earth Planet Sci Lett 80(3–4):252–264

References

Peters T (2005) Blatt St. Moritz 1:25,000, Geologischer Atlas der Schweiz #118, 1257

Peters T, Dietrich VJ, Ziegler WH, Nievergelt P, Pauli C (2008) Blatt Bivio 1:25,000, Geologischer Atlas der Schweiz #124, 1256

Schmutz HU (1976) Der Mafitit-Ultramafitit-Komplex zwischen Chiavenna und Val Bondasca (Provinz Sondrio, Italien; Kt. Graubuenden, Schweiz). Beitr Geol Karte Schweiz, Neue Folge 149, 73 pp

Schmid SM, Froitzheim N (1993) Oblique slip and block rotation along the Engadine Line. Eclogae Geol Helv 86(2):569–593

Spillmann P, Trommsdorff V (2005) Blatt Piz Bernina 1:25,000, Geologischer Atlas der Schweiz #119, 1277

Studer B (1851) Geologie der Schweiz (Band 1). Stämpfli, Bern/Schulthess, Zürich, 485 pp

Trommsdorff V, Montrasio A, Hermann J, Müntener O, Spillmann P, Giere R (2005) The geological map of Val Malenco. Schweiz Mineral Petrogr Mitt 85:1–13

Vasin R, Kern H, Lokajicek T, Svitek T, Lehmann E, Mannes DC, Chaouche M, Wenk H-R (2017) Elastic anisotropy of Tambo gneiss from Promontogno, Switzerland: a comparison of crystal orientation and microstructure-based modeling and experimental measurements. Geophys J Int 209:1–20

Weibel M, Locher T (1964) Die Kontaktgesteine im Albigna- und Fornostollen (nördliches Bergeller Massiv). Schweiz Mineral Petrogr Mitt 44:157–185

Wenk H-R (1980) More porphyritic dikes in the Bergell Alps. Schweiz Mineral Petrogr Mitt 60:145–152ADD

Wenk H-R (1983) Mullite-sillimanite intergrowth from pelitic inclusions in Bergell tonalite. N Jb Mineral Mh 146:1–14

Wenk H-R, Cornelius S (1977) Blatt Sciora 1:25,000, Geologischer Atlas der Schweiz #70, 1296

List of Geological Maps with Itineraries

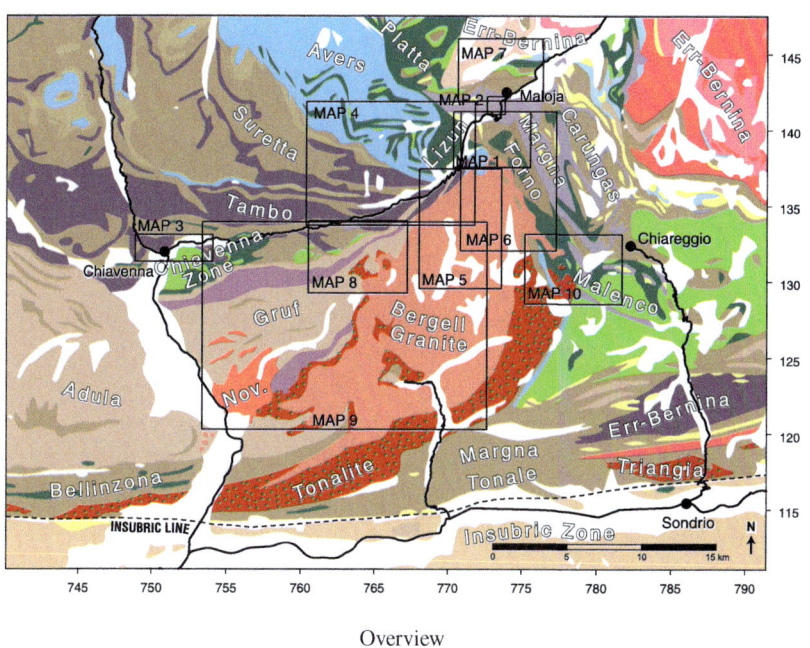

Overview

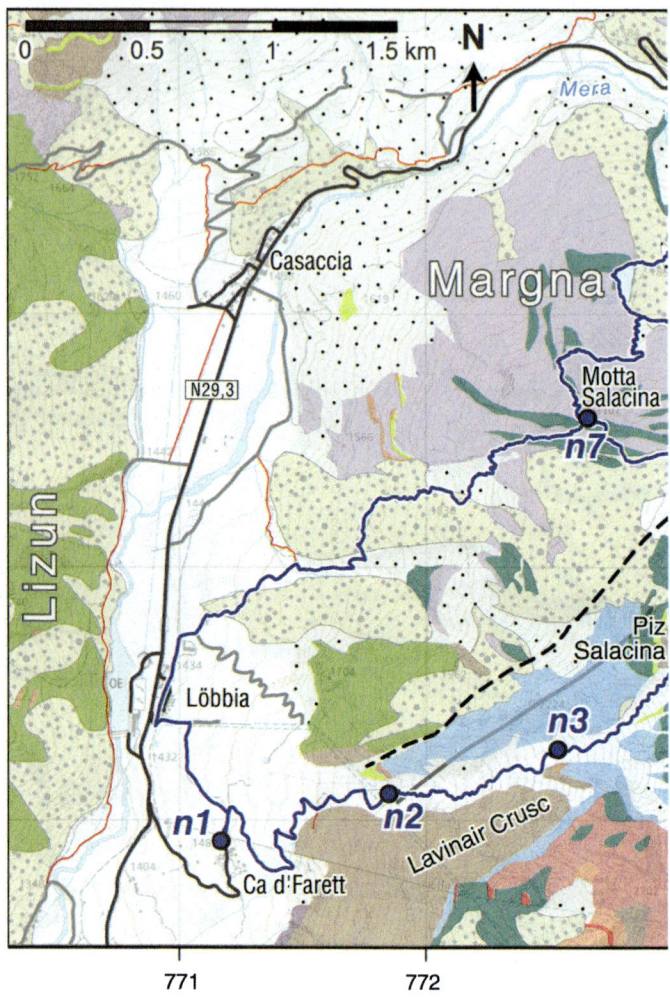

Map 1 Maloja-Cavloc-Casaccia-Piz Salacina

List of Geological Maps with Itineraries

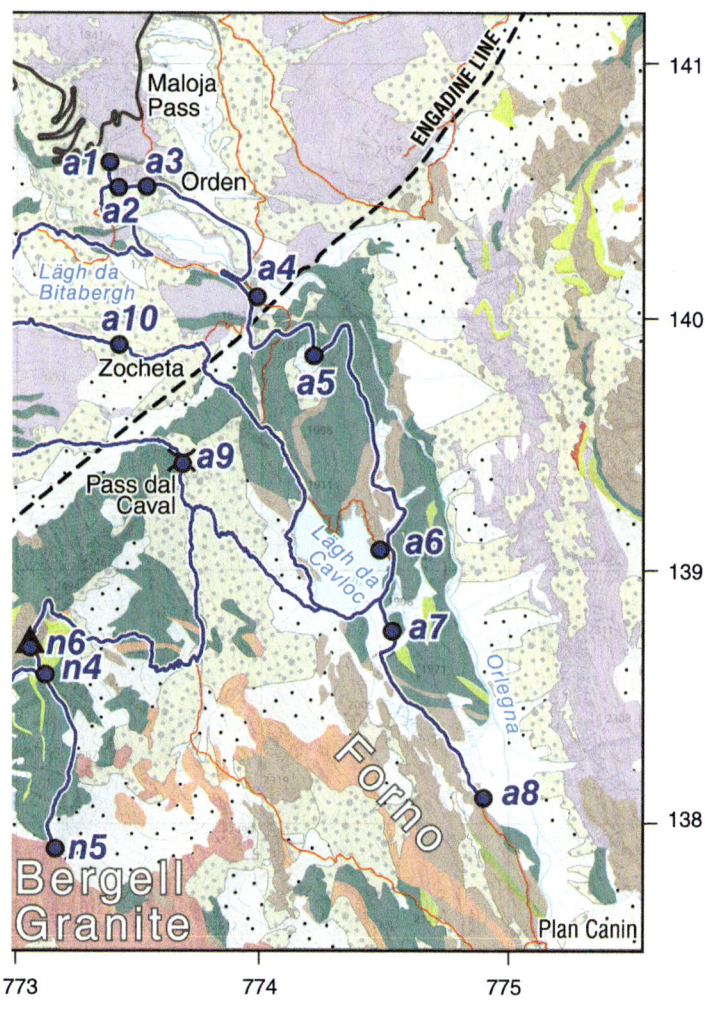

(continued)

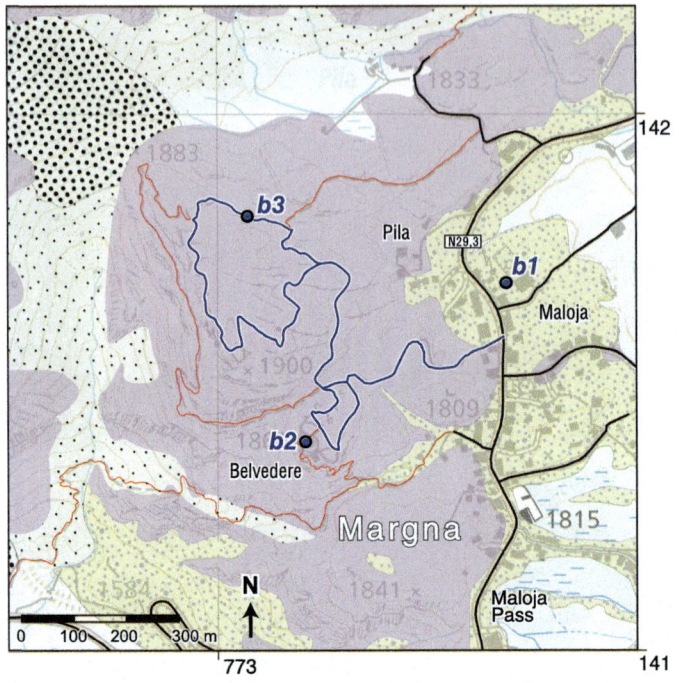

Map 2 Maloja-Torre Belvedere-Glacial mills

List of Geological Maps with Itineraries

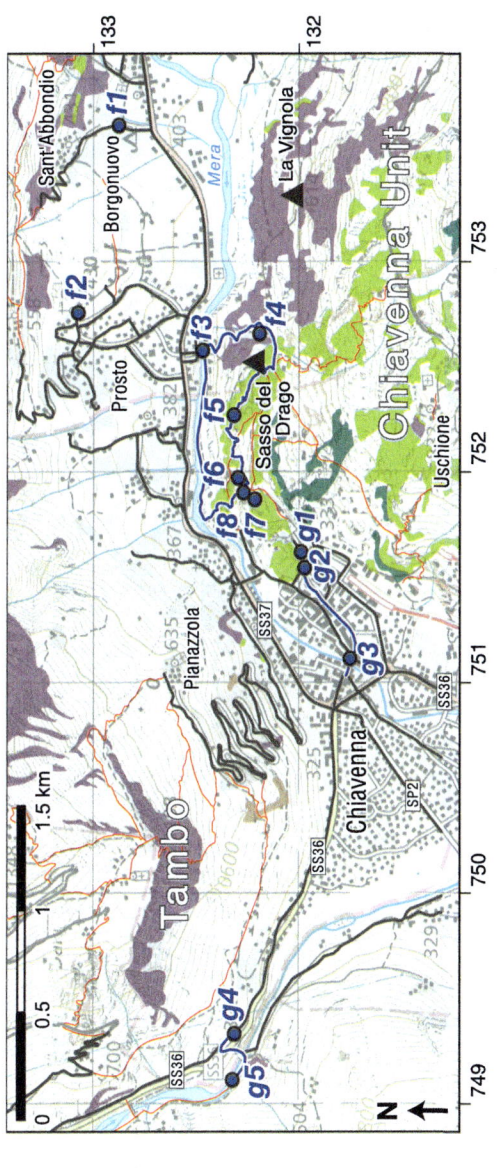

Map 3 Piuro-Chiavenna

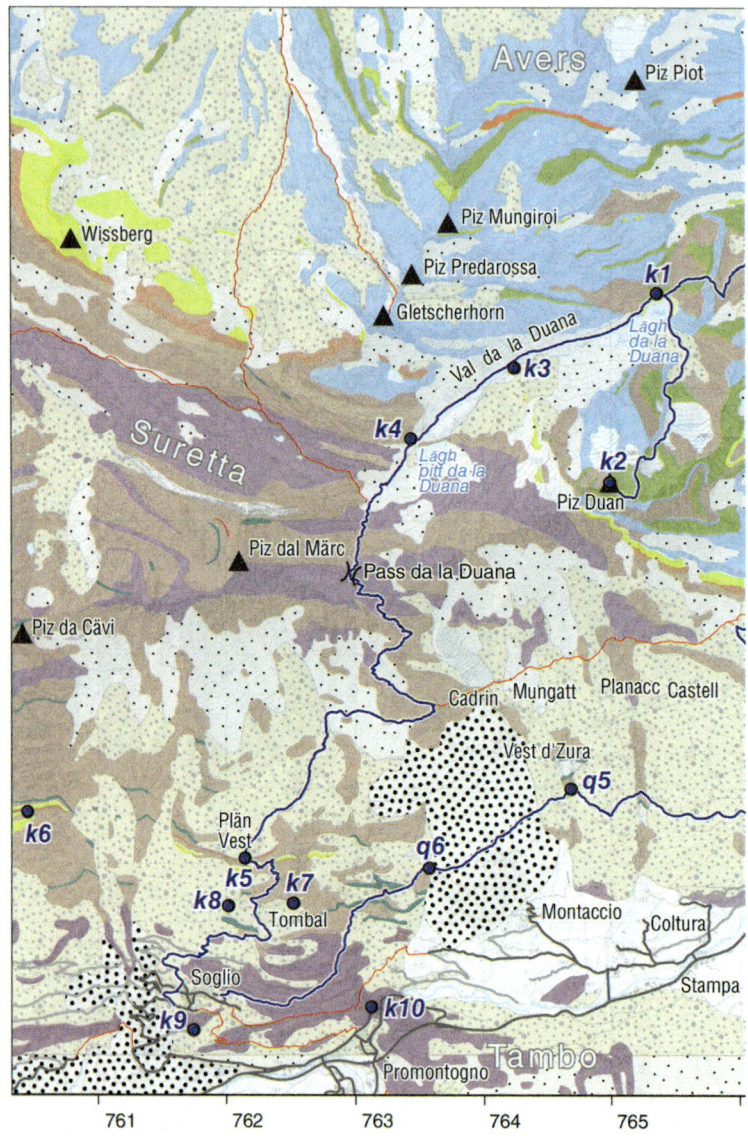

Map 4 Casaccia-Val Maroz-Van Duana-Soglio

List of Geological Maps with Itineraries 131

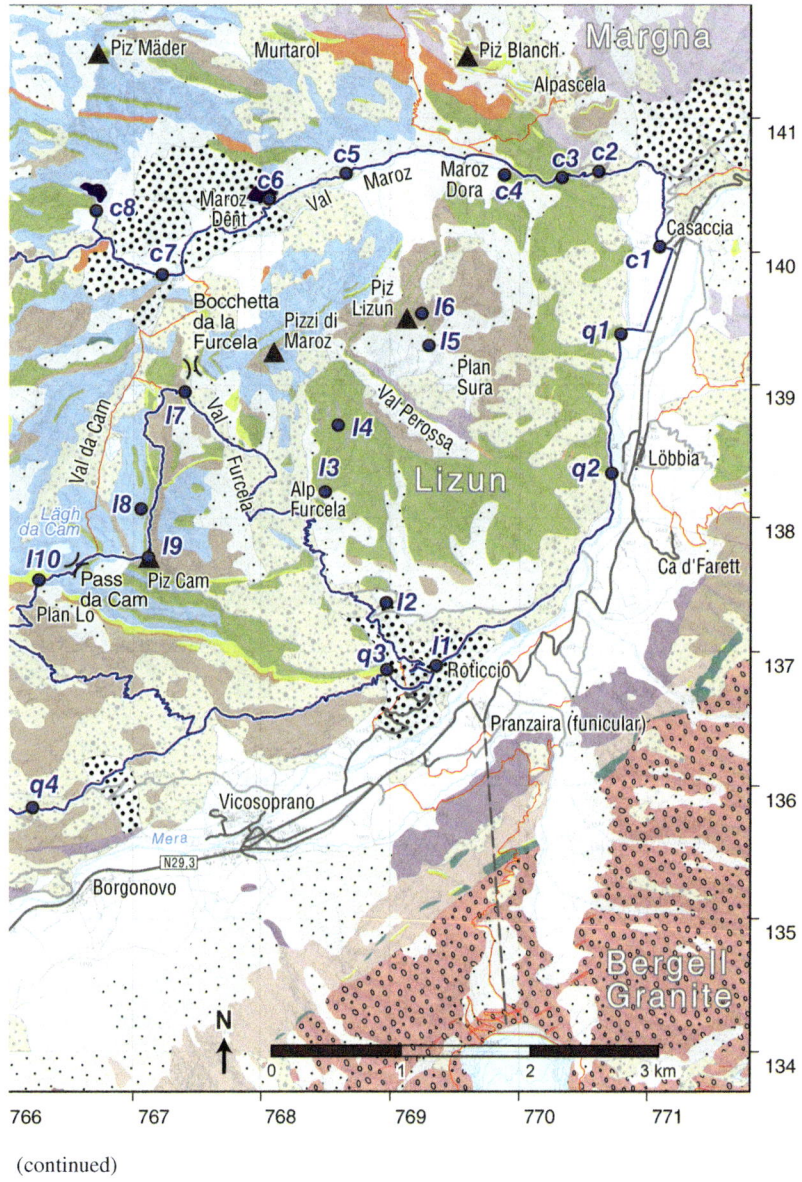

(continued)

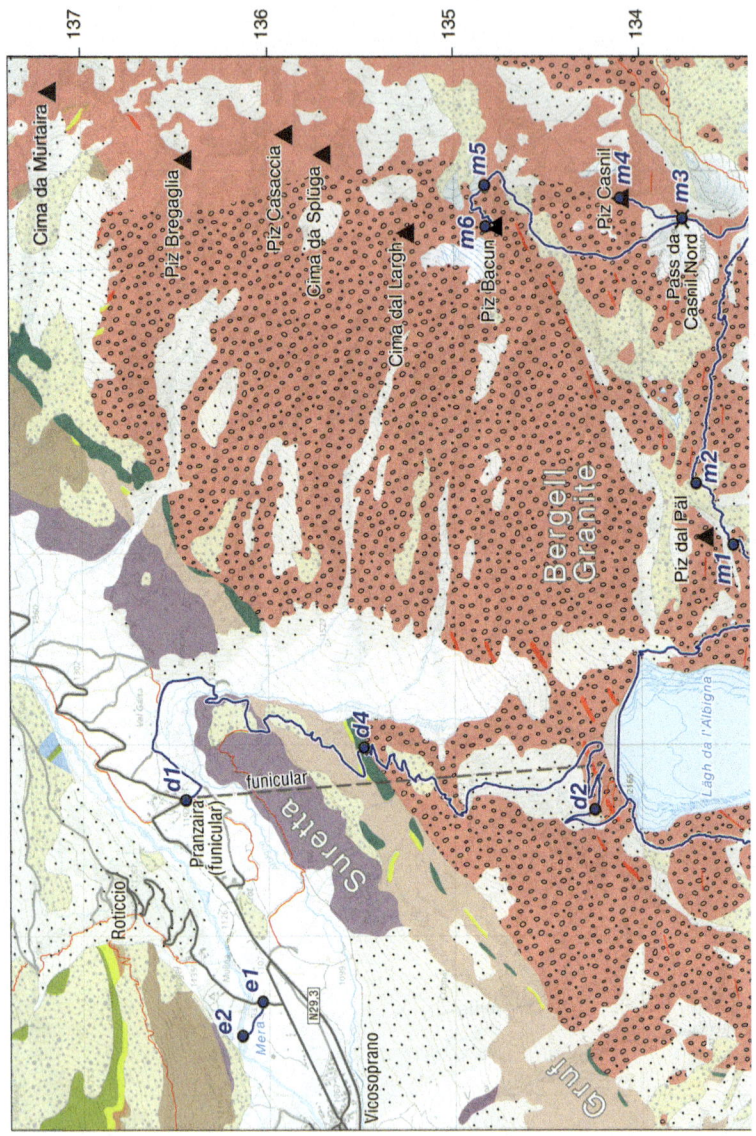

Map 5 Val Albigna

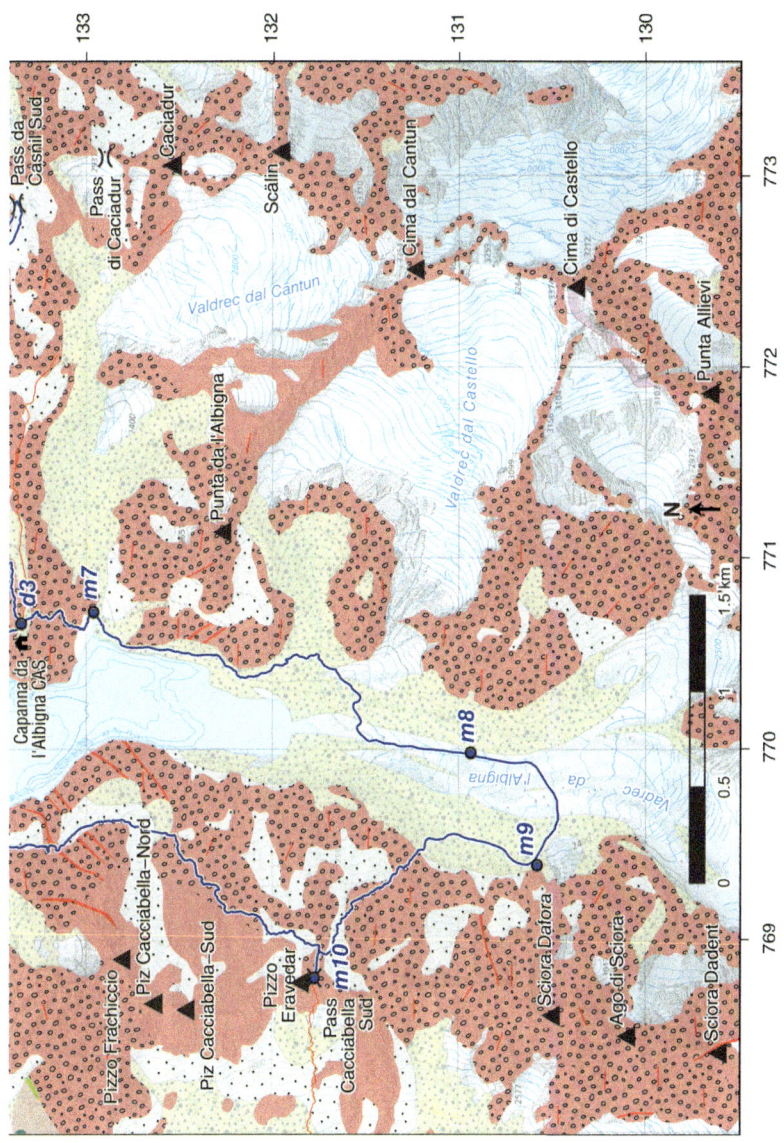

(continued)

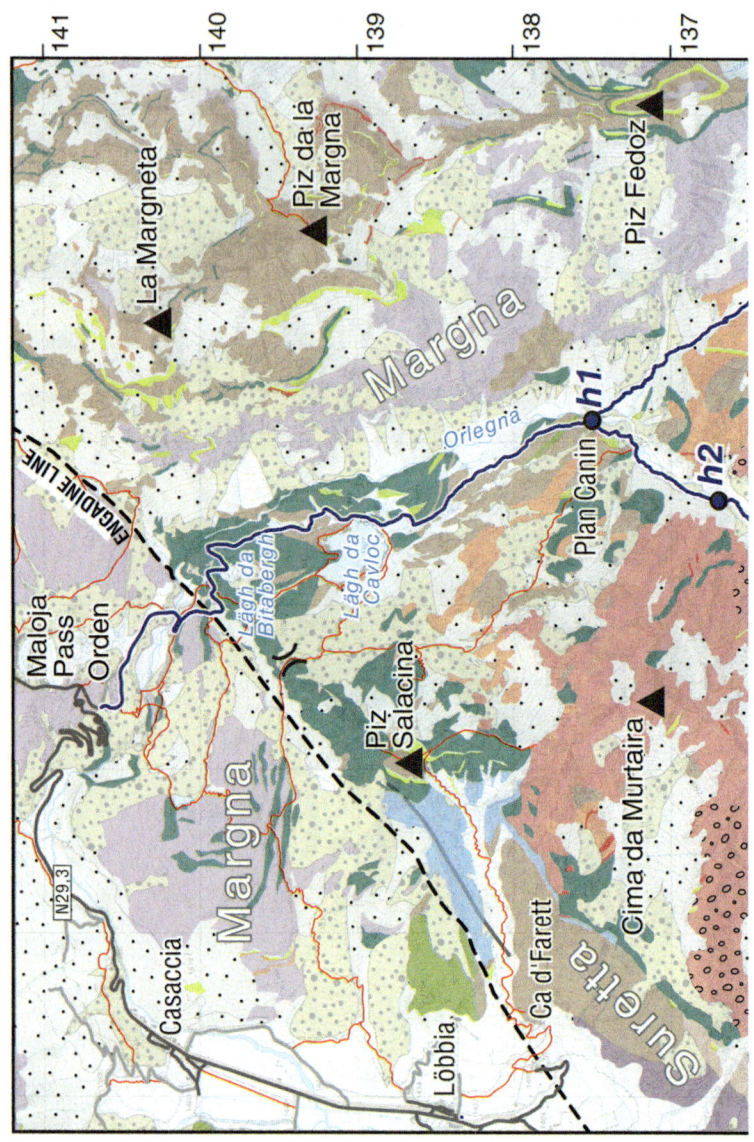

Map 6 Val Forno

Glossary

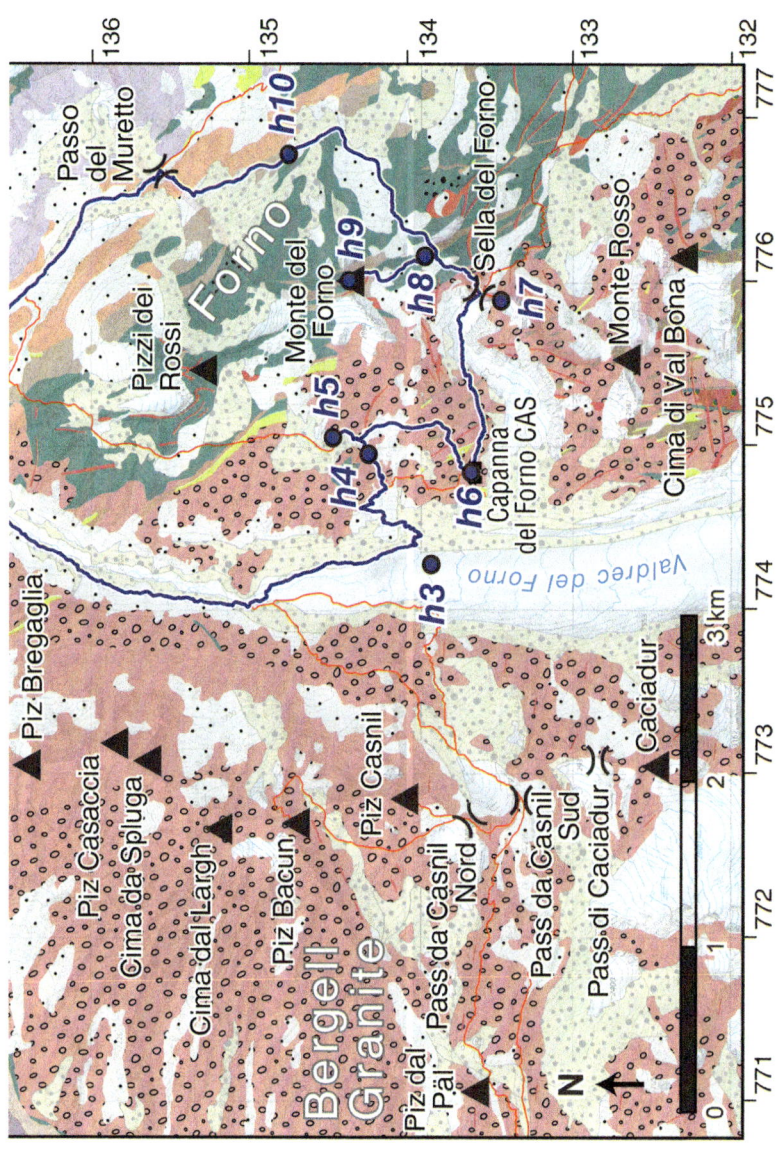

(continued)

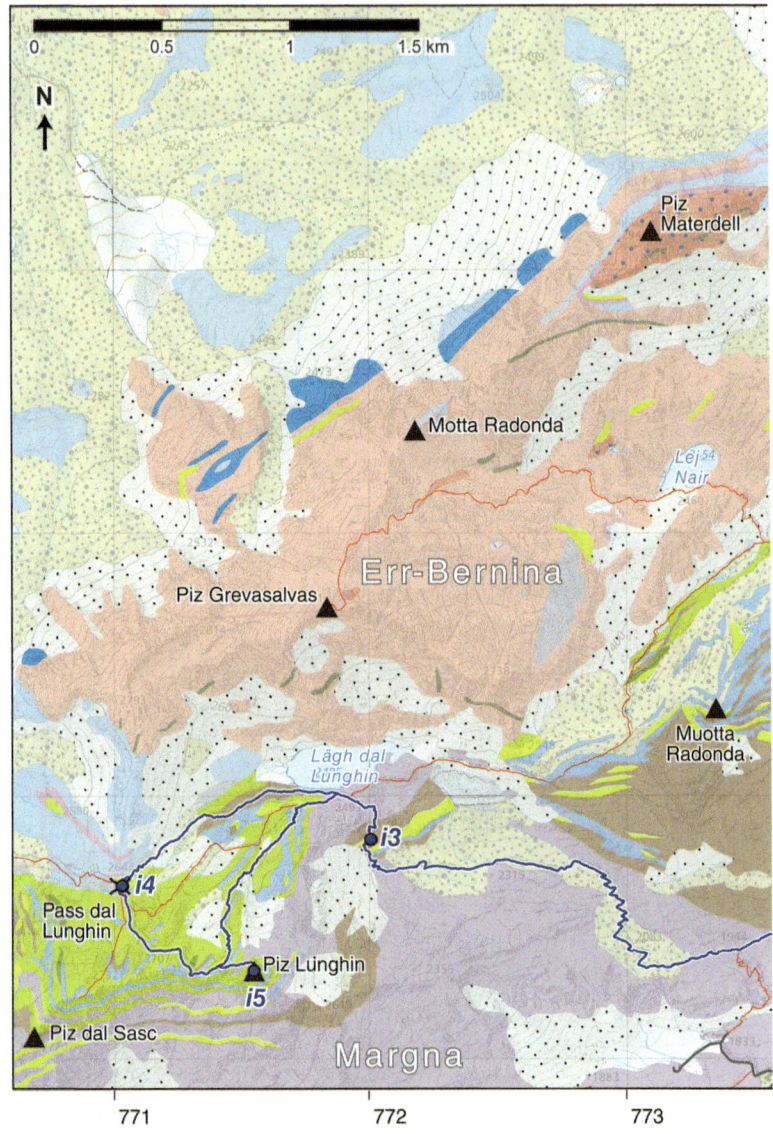

Map 7 Piz Lunghin, Grevasalvas-Piz d'Emmat

List of Geological Maps with Itineraries 137

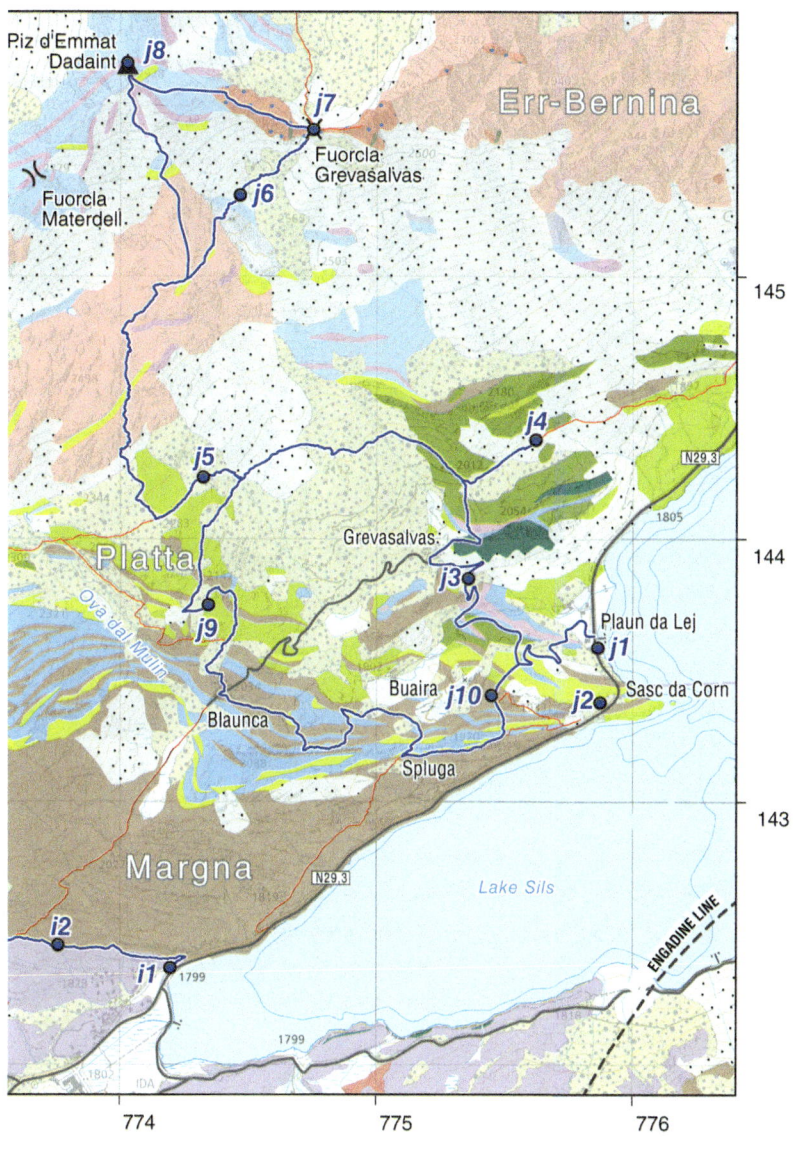

(continued)

List of Geological Maps with Itineraries

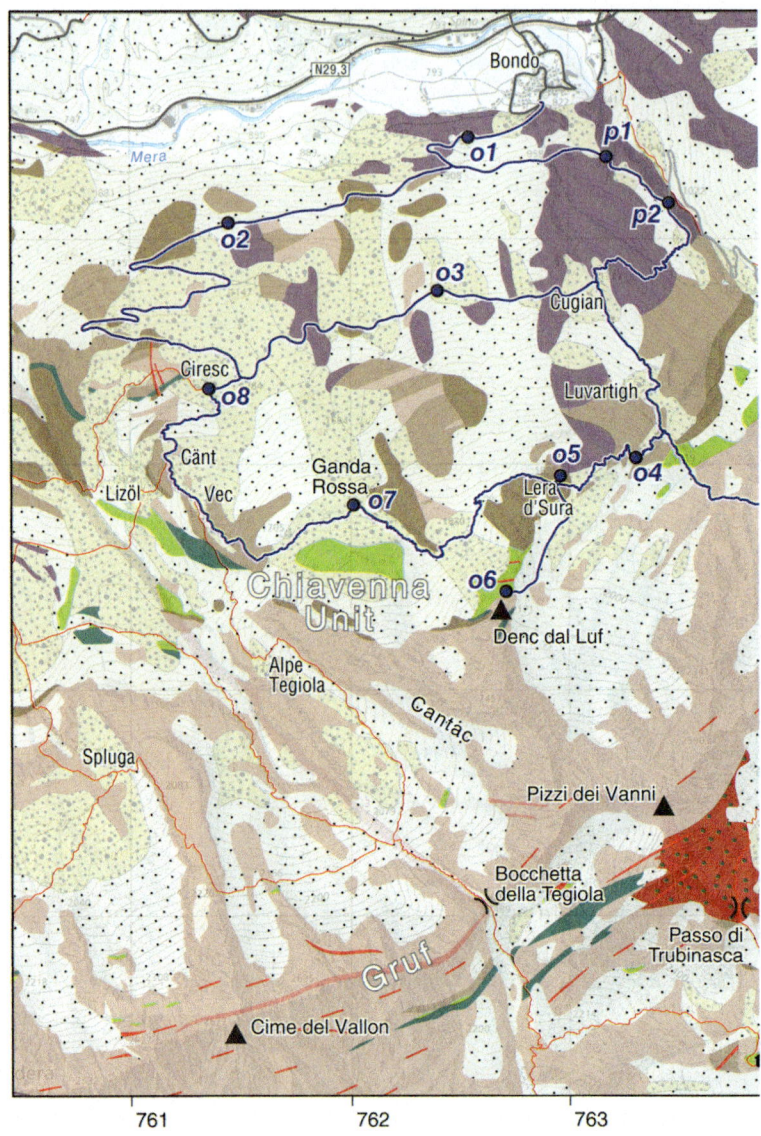

Map 8 Val Bondasca

List of Geological Maps with Itineraries

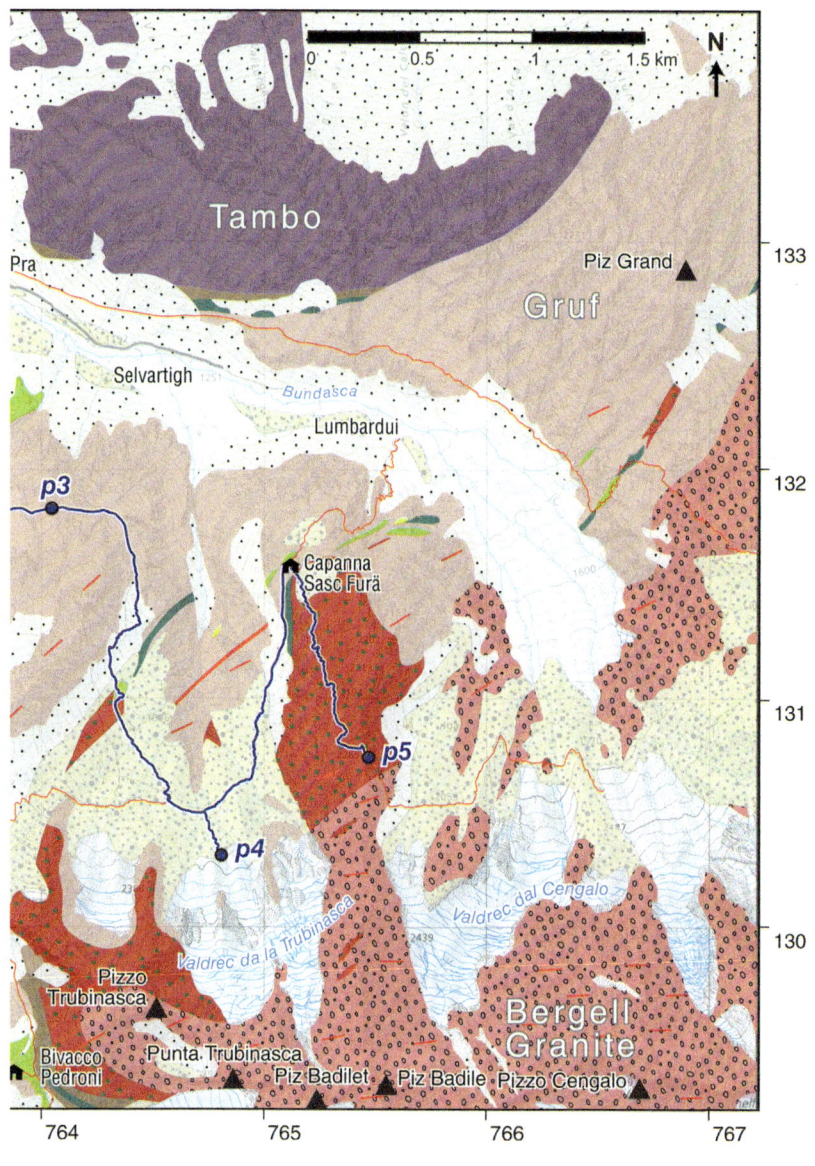

(continued)

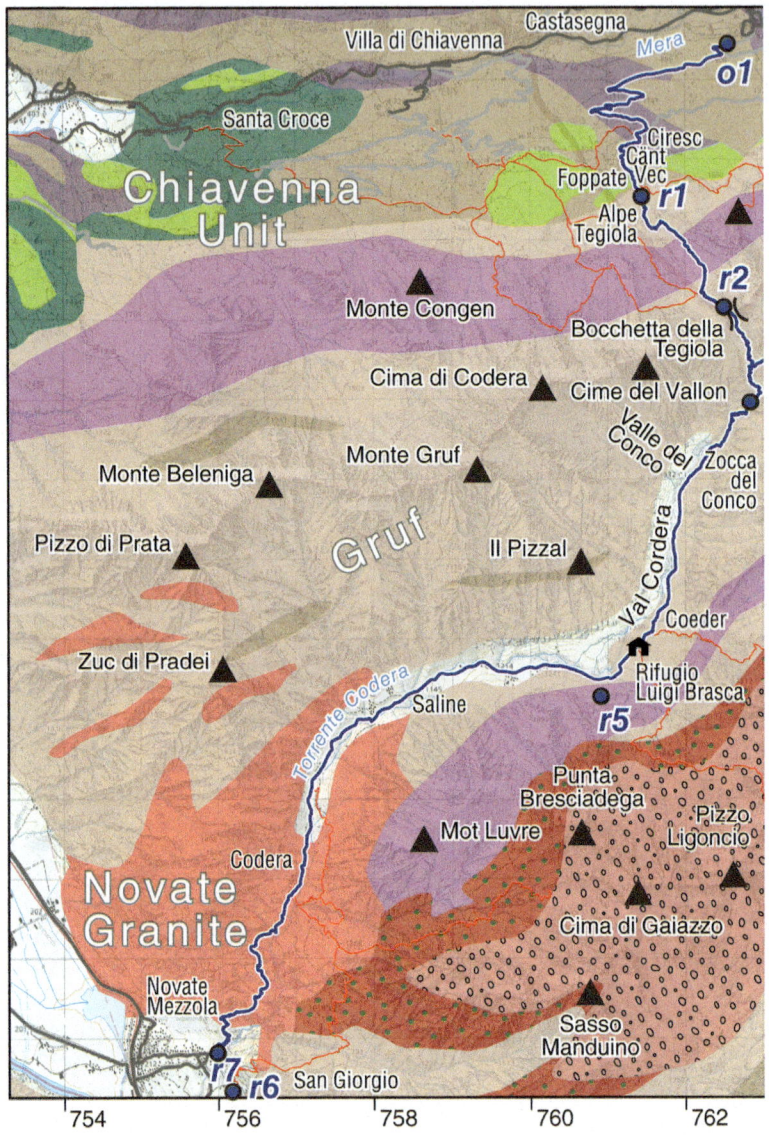

Map 9 Val Codera

List of Geological Maps with Itineraries

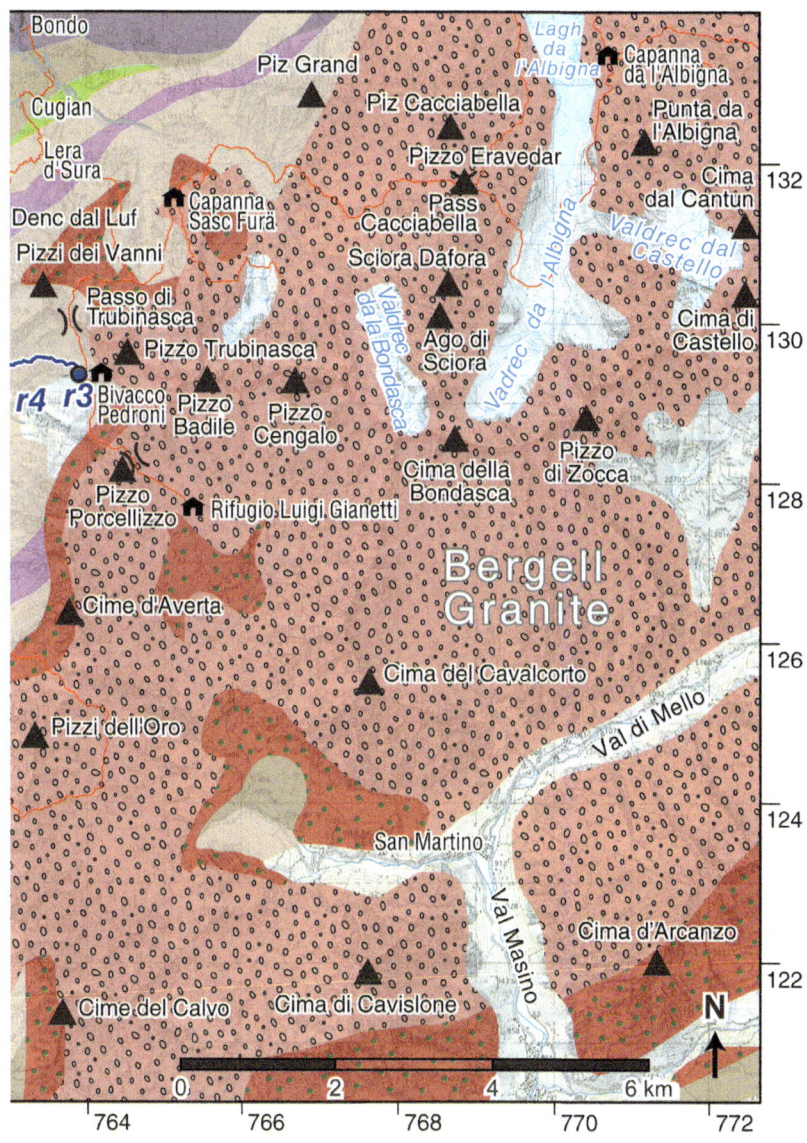

(continued)

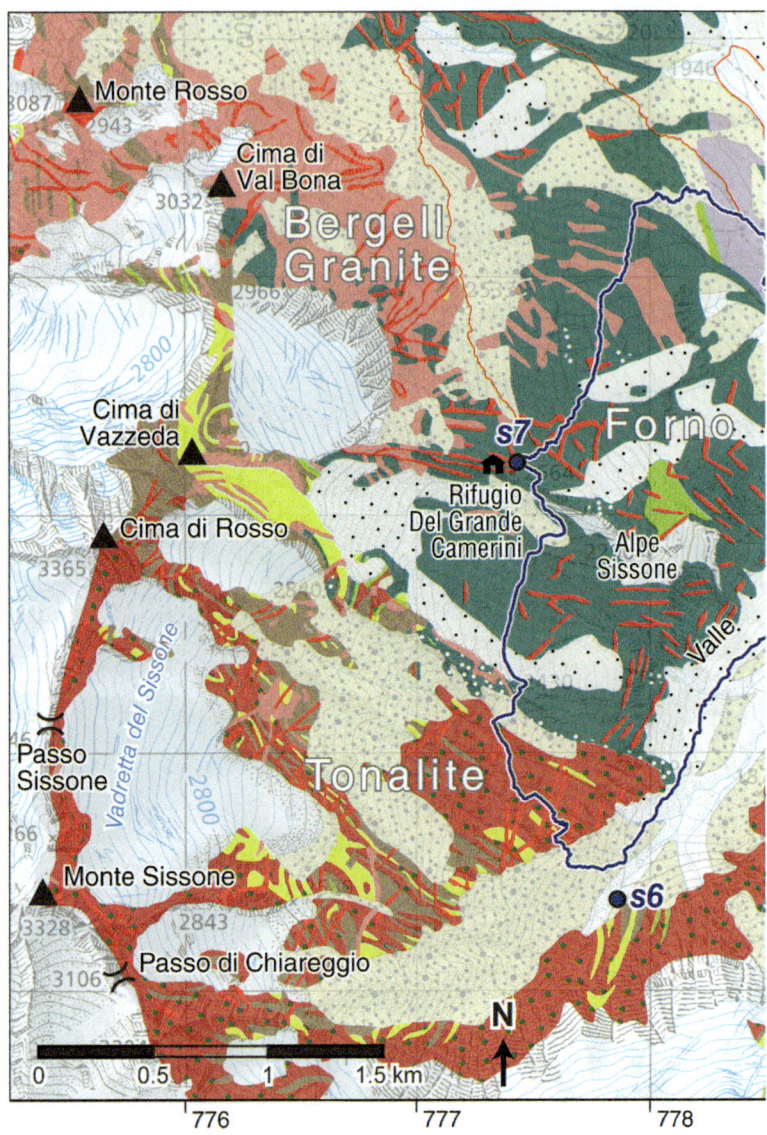

Map 10 Val Schiesone (Malenco)

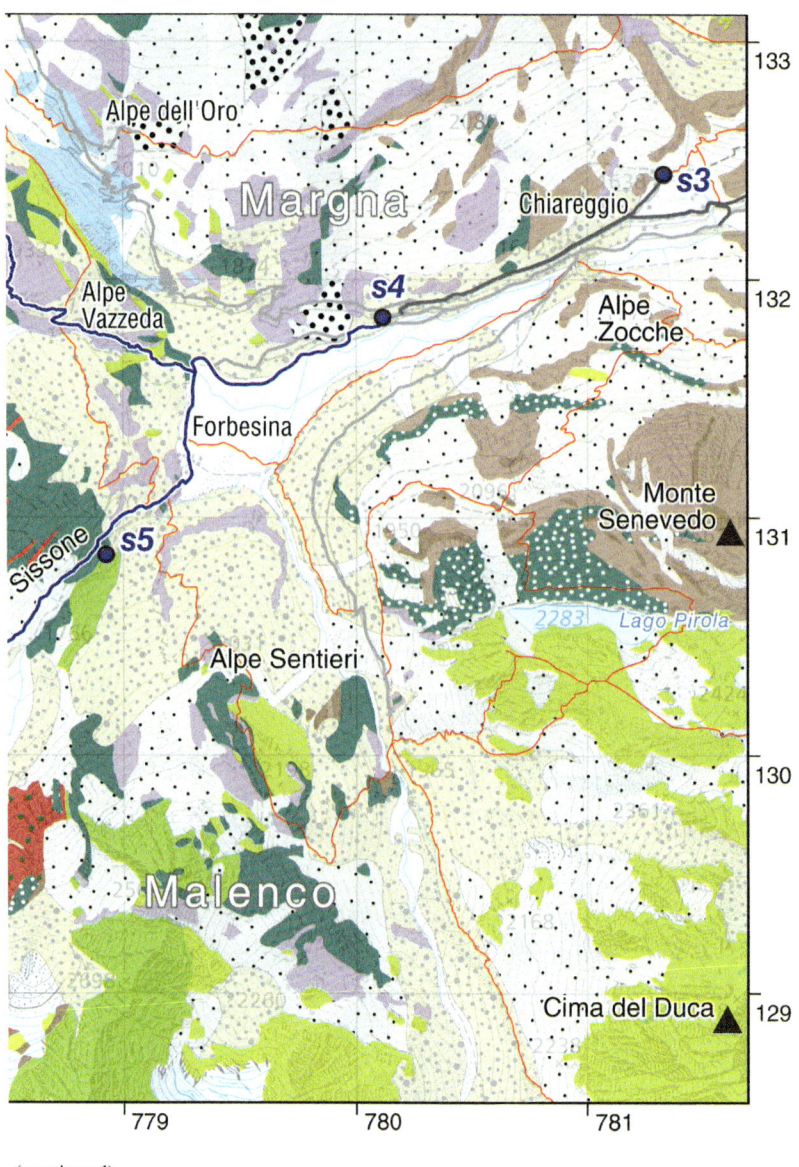

(continued)

Legend Geological Maps

Quaternary
- Alluvium
- Talus
- Rockfalls / Landslides
- Moraine
- Travertine

Tertiary Intrusions
- Bergell granodiorite, megacrystic
- Bergell granodiorite, homogeneous
- Tonalite
- Novate granite
- Pegmatite and aplite dikes
- Diabase dikes

Pennine nappes
Gruf Complex
- Gruf migmatites
- Orthogneiss

Tambo Nappe
- Orthogneiss
- Paragneiss
- Triassic quartzite
- Triassic marble

Suretta Nappe
- Orthogneiss
- Paragneiss
- Graphite schist
- Triassic quartzite
- Triassic marble

Avers Nappe
- Bündnerschiefer (Calcaceous schist)
- Greenschist (Prasinite)
- Glaucophane schist (Paragneiss)
- Triassic marble

Symbols
- Summits ▲
- Alpine huts 🏠
- Geologic waypoints ●
- Itineraries
- Other trails

Meta-oceanic Units
Chiavenna Zone
- Olivine-talc schists (Lavez)
- Amphibolite
- Paragneiss

Malenco zone
- Serpentinite
- Olivine-talc schists (Lavez)
- Amphibolite
- Gabbro
- Paragneiss

Forno zone
- Amphibolite
- Amphibolite with pillow structures
- Andalusite schist (Paragneiss)
- Diopside schist
- Calcitic schist

Lizun Unit
- Greenschist
- Serpentinite
- Orthogneiss
- Paragneiss
- Quartzite

Platta Nappe
- Serpentinite
- Greenschist
- Paragneiss

Austroalpine Nappes
Margna Nappe
- Orthogneiss
- Paragneiss
- Triassic marble
- Amphibolite

Err Nappe
- Granodiorite
- Tonalite
- Diabase dikes
- Orthogneiss
- Triassic marble
- Agnelli liassic siliceous limestone
- Jurassic radiolarite
- Cretaceous limestone

Legend

Glossary

Adamello batholith with various intrusions of ~90,000 km^3 in volume occurred mainly 43–40 My. It is located about 75 km E and 15 km S of the Bergell granite.

Alluvium is loose sand, gravel and larger boulders deposited by rivers.

Amphibolite is a metamorphic rock of mafic composition (hornblende, plagioclase, biotite) and transformed from basalt at relatively high temperatures.

Austroalpine Austroalpine nappes are a stack of nappes of Adriatic/African/Gondwana origin. They were thrust on top of European Pennine nappes.

Basalt is a mafic volcanic rock typical of oceanic crust and extruding along midoceanic ridges. There is no basalt in the Bergell Alps but a lot of amphibolite that was transformed from basalt during Alpine metamorphism. Occasionally pillow structures display the origin.

Bergell granite a Tertiary intrusion mainly of granodioritic composition on the southside of Val Bregaglia (35 My).

Bernina granite is an old Hercynian granite (333 My) in the Austroalpine nappes.

Boudinage from French "string of sausages" describes metamorphic rocks deformed in compression/shear where hard layers break up into pieces and soft layers intrude into gaps.

Bündnerschiefer is a low-grade metamorphic rock transformed from calcareous shales and slates and is of Jurassic age. It is named after the Swiss canton Graubünden.

Charnockite is an orthopyroxene-bearing quartz-feldspar metamorphic rock formed at extreme temperatures and pressures (granulite facies). It is named after the tombstone of Job Charnock in Kolkata (India) which is made of charnockite.

Contact rocks are rocks next to plutonic intrusions. They are often recrystallized with high temperature mineral assemblages.

Cornelius Hans Peter, 1888–1950, geologist, Wien. Ph.D. at ETH.

Dikes are igneous intrusions along fractures. Siliceous dikes in the Bergell are pegmatite and aplite, mafic dikes are diabase. Pegmatite dikes are coarse-grained, often zoned and contain large crystals of minerals like beryl and garnet; aplite has a smaller grain size.

Engadine Line left lateral fault on the S-side of Engadine Valley.

Giacometti Alberto, 1901–1966, artist, grew up in Stampa, later lived in Paris.

Giacometti Augusto, 1877–1947, painter, born in Stampa.

Gneiss is a metamorphic rock composed mainly of feldspar, quartz and mica. Orthogneiss refers to gneiss of originally granitic composition and paragneiss to sedimentary composition. Augengneiss is orthogneiss with large alkali feldspars ("Augen" German for eye). Examples are in the Tambo, Suretta and Margna nappes.

Gondwana A supercontinent and southern part of Pangea containing present day S-America, Africa, Antarctica, Australia, Arabia and India.

Granite is an igneous-plutonic rock composed largely of plagioclase, alkali feldspar and quartz with minor biotite, muscovite and hornblende.

Greenschists are mafic metamorphic rocks (rich in Mg) that recrystallized at low temperature and pressure conditions. Originally they were rocks like basalt and amphibolite. They are composed of albite, chlorite, actinolite and epidote.

Hardness is a property used to identify minerals. There is the Mohs hardness scale from 1 to 10 with 1 softest (e.g. Talc) and 10 highest (e.g. diamond). Typical are gypsum 2, calcite 3, quartz 7. Thumbnail is 2, knife blade 5, porcelain chip 6.

Ice age The last Ice Age (Riss-Würm) was 115,000–10,000 years before present (ybp), with a last glacial maximum ~25,000 ybp. A Little ice age occurred 1300–1500 ad, with cold temperature after a Roman warm period.

Insubric (Tonale, Canavese) Line A young (< 5 My) right lateral fault in the southern Alps extending over >500 km from Southern Austria (E) to NW Italy (W).

Kiln a lime kiln is an oven built with rocks to calcinate limestone and produce lime CaO. Required temperatures are 900–1000 °C.

Laurasia Northern part of Pangea, containing N-America, Greenland, Europe, Siberia. It was separated from Gondwana by the Tethys Ocean.

Lavez (also known as laveggio, soapstone, pietra ollare or talc-olivine-chlorite schist) is an ultramafic rock with unique properties that make it suitable for the production of pots and statues.

Lime CaO was used for construction of buildings like modern Portland cement by hydrating it to produce $Ca(OH)_2$ which subsequently transformed to $CaCO_3$ by carbonation.

Mafic rocks are relatively poor in silica and alkali feldspars. Examples are basalt, amphibolite and greenschist.

Marble was originally limestone that has recrystallized during metamorphism. Mineral composition is either calcite or dolomite.

Metamorphic facies are based on mineral assemblages typical of temperature–pressure conditions. Typical facies in the Bergell Alps are Greenschist facies (low temperature, low pressure), Amphibolite facies (intermediate temperature and pressure), Granulite facies (high temperature, close to melting of granite) and some Blueschist facies (low temperature intermediate pressure).

Metamorphic rocks transformed from igneous rocks (e.g. basalt, granite), sedimentary rocks (such as limestone, sandstone) or old metamorphic rocks by recrystallization at different temperature–pressure environments and often expressing significant deformation. Typical metamorphic rocks are gneiss, schist, amphibolite, marble and quartzite.

Migmatites are high temperature metamorphic rocks of granitic composition, where gneiss undergoes partial melting.

Minerals Naturally occurring chemically homogeneous compounds, most of them crystalline with a regular periodic lattice structure and arrangement of atoms.

Moraines are material of debris (clay, sand, rocks) transported and deposited by glaciers. There are lateral moraines on the side of glaciers, terminal moraines at the end of a glacier and basal moraines at the base of glaciers. Because of high concentrations of clays, moraines are generally very fertile.

Mylonites are metamorphic rocks that underwent extreme plastic deformation, sometimes in layers with a fine-grained matrix and often some larger inclusions.

Nappe Tectonic units that were thrust over each other during Alpine tectonic events and the collision of Laurasia and Gondwana.

Novate granite also known as San Fedelino granite is the youngest granite in the Bergell Alps (20–25 My).

Pangaea Giant continent that covered the Earth from late Paleozoic (300 My) to late Triassic (240 My).

Pelitic schists are metamorphic rocks transforming from argillaceous (Al-rich) sediments such as shales. They may contain aluminosilicates such as andalusite, kyanite, sillimanite indicative of temperature–pressure conditions.

Pennine Pennine nappes are a stack of sheets with European origin juxtaposed during Alpine folding.

Petrographic microscopes are used by geologists to study thin sections of rocks with polarized light to identify minerals and microstructures.

Pillow basalt Basalt has a very low viscosity and can flow over large distances but if it extrudes into water it is quenched and forms spherical-shaped bodies which are called pillows. Near Muretto Pass such spherical structures are observed in amphibolite, reminding us that they represent oceanic crust that has been subjected to metamorphism.

Plate tectonics Tectonic plates are composed of oceanic lithosphere and thicker continental lithosphere that are moving at velocities of ~2 cm per year, joining and dividing over geologic times. The movement is driven by convection in the Earth's mantle, including subduction and uplift.

Platta nappe The Platta nappe is of particular interest in the Bergell region because it represents oceanic crust of the Tethys ocean and is juxtaposed between Pennine and Austroalpine nappes.

Plurs-Piuro is a town in Valle della Mera that was devastated by the 1618 landslide where over 2000 people were killed.

Plutonic rocks crystallized from a melt at depth and have generally large grain size. Phenocrysts such as alkali feldspar in Bergell granite are large euhedral crystals that were growing in the melt. Plagioclase may display changes in chemical composition (zoning) indicating temperature changes during crystallization.

Prasinite is a low-grade metamorphic rock of mafic composition such as basalt. The name comes from Greek "prasinos", leek-green. Typical minerals are albite, chlorite, epidote, actinolite.

Radiolaria are protozoa typically 0.1–0.2 mm in size with intricate often siliceous mineral skeletons. The earliest radiolaria have been documented in Cambrian. In the Bergell they appear to be Jurassic.

Recrystallization If rocks are subjected to high temperature and pressure new crystals form by nucleation and grain growth. All metamorphic rocks recrystallized from previous igneous or sedimentary rocks.

Rocks are composed of minerals and form at a wide range of temperature–pressure conditions. Sedimentary rocks form close to the surface, plutonic rocks at depth. They may be transformed to metamorphic rocks during the geological history.

Sedimentary rocks form near the surface of the Earth through compaction and lithification of sediments in aqueous environments. Examples are limestone, sandstone and shale.

Segantini Giovanni, 1858–1899, Italian painter who lived in Maloja.

Serpentinite is a metamorphic rock of ultramafic composition (mainly Mg, low Si; largely mineral serpentine). It transformed from peridotite (olivine, pyroxene), mainly from upper mantle, by hydration.

Solid solution Many minerals can have a range in composition without changing the structure significantly. Examples are olivine (Mg–Fe) and plagioclase feldspar (Na, Si–Ca, Al).

Staub Rudolf, 1890–1961, geologist, ETH Zurich.

Studer Bernhard, 1794–1887, geologist, Univ. Bern.

Talus is a slope formed by accumulation of rock debris below cliffs.

Tectonic units The Earth's lithosphere is divided into tectonic units. This includes large plates (such as Eurasia) or smaller units such as thrust sheets (or nappes) that have been juxtaposed in the Alps (such as Tambo, Margna nappe in the Bergell).

Tethys Ocean formed in late Triassic (~ 250 My) when the supercontinent Pangea started dividing into Laurasia (N) and Gondwana (S).

Time Geological time has been divided into periods and episodes over 4.5 By.

Tonalite is a mafic granitic rock composed mainly of plagioclase, quartz, hornblende, biotite and some alkali feldspar. It is named after Passo del Tonale and the type locality is part of the Adamello intrusion.

Travertine is a form of porous limestone deposited around mineral springs. Famous travertine deposits are in Yellowstone (USA), Pamukkale (Turkey) and was mined in Tivoli near Rome for over 2000 years. You may discover some travertine in Val Maroz.

Triple point If three fields of particular properties meet on a two-dimensional surface they define a point. This can be water drainage such as North-Sea, Adriatic, Black Sea, or mineral polymorphs such as Al_2SiO_5 on a pressure–temperature phase diagram.

Ultramafic rocks are depleted in silica, aluminum, potassium and sodium. The main constituents are Mg, Fe, some Si. Typical ultramafic minerals are olivine, talc, serpentine, chlorite and diopside. Serpentinite, peridotite and lavez are ultramafic rocks.

Volcanic rocks form when melt reaches the surface, produces lava flows and gets quenched at surface temperatures. They may contain glass and often inclusions of minerals that crystallized in the melt at depth (phenocrysts).

Xenolith from Greek xeno (foreign) and lithos (stone) is a dark, hornblende-rich inclusion in granite, often occurring as "swarms".

Index

A
Accommodations, 68
  Cap. Albigna SAC, 77, 104
  Cap. Forno SAC, 89
  Cap. Sasc Furä SAC, 112
  Cap. Sciora SAC, 105
  Biv. Pedroni Del Prà, 117
  Rif. Del Grande Camerini CAI, 121
Actinolite, 26
Adamello granite, 20
Ago di Sciora, 110
Albigna, 102–105
Albigna, Cap. SAC, 77, 104
Albigna dam, 60, 76
Albite, 27, 40, 71
Alkali feldspar phase diagram, 39
Alluvium, 35, 145
Alpe Tegiola, 116
Alteration of granite, 103
Aluminosilicates, 43
Amphibolite, 11, 12, 33, 71, 76, 79, 86, 90, 106, 115, 145
Amphibolite facies, 14, 16, 71
Andalusite, 4, 25, 41, 43, 71, 86, 91, 109
Anorthite, 27, 40, 71
Anthropogenic global warming, 45
Antigorite, 26, 33, 92, 119
Aplite, 31, 76, 89, 109
APQ diagram, 37, 39
Apulian microcontinent, 11
Aquafraggia waterfall, 80

Augengneiss, 69, 85, 91, 97, 114–115
Austroalpine nappes, 12, 13, 19, 93, 95, 107, 119, 145, 148
Avers nappe, 11, 19, 75, 97, 101, 102
Azurite, 24

B
Bacun, Piz, 104
Badile, Pizzo, 38, 40, 109, 110, 112
Basalt, 12, 30, 89
Bergell granite, 3, 11, 31, 37, 39, 76, 102–104
Bergell granite age, 8, 31
Bergell intrusion, 14
Bernina nappe, 11
Bernina, Piz, 19
Bernina granite age, 8
Beryl, 25, 28, 104, 116
Biotite, 26
Bitabergh Lake, 72
Bivacco Pedroni Del Pra, 116
Blaunca, 96
Blueschist facies, 16, 95
Blueschists, 73, 95
Bondasca, Val, 51, 109, 110
Bornite, 57
Boudinage, 86, 87, 145
Bronze Age, slags, 57
Bündnerschiefer, 75, 97, 145

C

Cacciabella Sud, Passo, 104
Ca d'Faret, 106
Calcite, 24, 28
Camerini Del Grande, Rifugio, CAI, 121
Camping, 68
Cam, Piz, 65, 101
Cantun, Piz, 65
Carboniferous, 9
Casaccia, 3, 33, 35, 48, 57, 73, 96, 112
Casnil, Piz, 65, 104
Cävi, 98
Cavloc, Alp da, 71, 86
Cengalo, Pizzo, 3, 38, 40, 51, 105, 109
Chalcopyrite, 24, 56–58, 113, 121
Charnockite, 118, 145
Chiareggio, 119
Chiavenna, 81–83
Chiavenna zone, 53, 107
Chiavennite, 23, 27, 28, 30
Chiesa, 119
Chlorite, 26, 29, 76, 113
Chlorite-epidote schist, 76
Chloritoid, 25, 97, 99
Chrysotile, 26, 33, 53, 92
Ciäsa Granda Museum in Stampa, 1, 53
Cimaganda, 83
Ciresc, 107, 109
Clay, 35, 104
Cleavage of minerals, 29
Climate change, 45, 77, 88
Clinochlore (chlorite), 29
Clinohumite, 25, 88
Codera, Val, 4, 116
Colors of minerals, 30
Contact metamorphism, 121
Contact zone, 78, 88, 89, 106–107, 116, 121
Continental collision, 3, 11
Continental drift, 9
Coordinates, 67
Copper deposits, 56, 113
Cordierite, 15, 25, 116
Cornelius, H.P. (1888–1950), 18, 95, 116, 146
Corundum, 116

Cretaceous, 9, 16, 93
Cross-sections, 4, 16
  Alps, 12
  Bergell, 14, 17
  EWZ, 61
  Maloja (Cornelius), 18
  Maloja (Studer), 4

D

Dangers, 3, 65
Denc dal Luf, 109
Diabase dike, 31, 73, 91, 96, 100, 146
Diopside, 25, 88, 105, 149
Disgrazia, Monte, 91, 103, 119, 120, 121
Dolomite marble, 97
Dolomite, 24
Drescher-Kaden, Friedrich, 4, 111
Drosera, 35, 71, 96
Duan, Piz, 19, 65, 97
Dufour topographic map, 45–46
Durbegia, 115

E

Earth history, 8
Edelweiss, 97, 98
Electron microscope, 40
Elektrizitätswerke Zürich (EWZ), 59
  Cable car, 76, 102
  Albigna dam and lake, 76, 104
  Löbbia, 59–61, 113
Emmat Dadaint, Piz d', 96
Engadine, 92
Engadine line, 69, 146
Epidote, 25, 28, 116
Eravedar, Pizzo, 105
Erratic blocks, 45, 81
Err granodiorite, 91, 95
Err nappe, 11, 95
European nappes (Pennine), 12, 91, 93
Excursions, 65

F

Fedoz, Piz, 89, 107

Forno
 Capanna, SAC, 68, 69
 Copper deposits, 56
 glacier, 45–46, 66, 72, 86, 87
 Monte del, 89, 90
 Furcela, Alp, 100

G

Gabbro, 11, 33, 96
Ganda Rossa landslide, 48, 109
Garnet, 15, 25, 28, 31, 38, 41, 71, 79, 88, 97, 106, 116, 118, 121
Geological maps, 65, 125–144
Ghiandone, 104
Giacometti Alberto, 1
Giacometti Augusto, 1, 55
Glacial mills, 3, 45, 47, 73, 81–82
Glacial morphology, 72, 93
Glacial polish, 45, 47, 77, 86, 102, 115
Glacial retreat, 45, 88
Glacier striations, 47, 69
Glaciers, 44–48, 59
Gneiss, 12, 32–34, 52, 56–57, 76, 79, 87, 89, 95, 100, 115, 146
Gondwana, 9, 93
Gonfolite, 47, 48
Granite, 31, 32, 146
 Homogeneous, 88, 102, 105, 110
 Megacrystalline, 37, 38, 102
Granite contact, 88, 107
Granodiorite, 37, 39, 76
Granulite facies, 16, 116
Gravel pit, 56, 113
Greenschist, 12, 33, 73, 97, 100, 113, 146
Greenschist facies, 16
Grevasalvas, Fuorcla, 95
Grevasalvas granite, 91
Grossular garnet, 88, 105, 106, 121
Gruf migmatite, 31, 107, 109, 110, 115–118
Gruf unit, 11, 107, 115
Gypsum, 24, 28, 99, 100

H

Hammer, 67
Hand specimen, 40
Hardness of minerals, 29, 146
Holocene, 9
Hornblende, 26, 29, 31, 33, 71, 76, 88, 96, 116, 121
Hydrothermal metamorphism, 38, 112
Hypersthene, 116

I

Ice, 36–37, 46, 48, 65–67, 69, 82, 97, 98
Ice Age, 44–48, 69, 146
Igneous rocks, 30, 32
Insubric/Tonale Line, 19, 119
Ion microprobe, 118
Isotope analyses, 7, 45, 118

J

Julier nappe, 11
Jurassic, 9, 75, 91, 97, 101

K

Karst structures, 97
Kutnahorite, 24, 30, 102
Kyanite, 4, 25, 41–43, 109

L

Lägh da la Duana, 97
Lägh pit da la Duana, 97
Lagrev, Piz, 92, 94–96
Landslides, 48–52
 Bondeno (2012), 83
 Casaccia (1673, 1970), 48, 50, 76, 113
 Cengalo/Bondasca (2017), 51, 109, 110
 Cimaganda (1533/34), 51, 83, 85
 Ganda Rossa, 109
 Gianda, 95
 Maroz, 48, 76, 96
 Piuro-Plurs (1618), 48, 49, 80
 Santuario di Gallivaggio (2018), 52, 86
 Val Torta (1988), 50

Last Glacial Maximum (LGM), 44
Laumontite, 27, 29, 102
Laurasia, 9, 93, 146
Laveggio, 52
Lavez, 52–54, 80–82, 121, 146
Lavez baptismal font, 85
Lavinair Crusc, 106
Lera d'Sura, 109
Ligoncio, Pizzo, 116, 117
Lime kiln, 55, 80, 93, 96, 98, 100, 146
Lime production, 53–55
Limestone, 9, 12
Liro river, 83–84
Little ice age, 45
Lizardite, 26, 29, 33, 92
Lizun, Piz, 100
Lizun unit, 73, 113
Löbbia power plant, 59, 113
Luigi Brasca, Rifugio, CAI, 68, 118
Lunghin Pass, 3, 44, 91
Lunghin, Piz, 92

M
Magnetite, 24, 75, 76
Maira river, 78
Malachite, 24, 28, 58
Malenco serpentinite, 119
Maloja, 69, 72–74, 91
Marble, 9, 12, 33–34, 52, 53, 75, 88, 91, 95–97, 99–102, 106, 115, 121, 147
Marble (Cima di Vazzeda), 88
Margna gneiss, 69, 70, 73
Margna nappe, 11, 18, 69, 89, 91, 93, 113
Margna, Piz da la, 91
Mariposite, 26, 29, 30
Marmitte dei Giganti, 82
Materdell, Piz, 96
Maroz, Val, 50, 65–67, 73–76, 96–97
Megacrysts, 31, 37–38, 70, 85
Melting point, granite, 16, 37
Metal ores, 56–58, 91
Metamorphic facies, 12, 16, 147
Metamorphic rocks, 30, 32–34, 147
Metasomatism, 110

Microcline, 27, 40
Microstructure of rocks, 40
Migmatites, 31, 109, 110, 118, 147
Mineral identification, 29, 30
Minerals, 23
Mining, 52–58
Moraines, 35–36, 44–48, 72, 100, 103–104, 113–115, 147
Mota Farun (Copper), 56–58
Motta Salacina, 72, 107
Mott Scalotta/Tavretga (Copper), 56–58
Mullite, 30, 121
Muretto Pass, 91
Murtaira, Piz, 106
Muscovite, 26, 27
Museums
    Ciäsa Granda, Stampa, 1, 28
    Museo Archeologico, Chiavenna, 2, 83
    Museo dei Minerali, Sondrio, 2, 28, 121
    Palazzo Castelmur, Stampa, 1
    Palazzo Vertemate, Franchi, Prosto, 1, 80, 81
Mylonite, 116

N
Nappes, 4, 11, 14, 18, 144, 147
    Avers, 19, 75, 97, 101
    Bernina, 16
    Err, 16
    Margna, 14, 16, 69, 89, 91, 113
    Platta, 11, 73, 89, 91, 93, 113
    Suretta, 14, 19
    Tambo, 14, 19
Novate granite, 8, 31, 118, 147
Novate-Mezzola, 118

O
Old granites, 12, 31, 37, 83, 93
Olivine, 25, 52, 109, 112, 121
Ophiolite zone, 89
Optical microscope, 40–41, 94, 99
Orbicular granites, 87, 89, 104
Ore minerals, 56, 88
Orlegna dam, 69

Orthoclase, 27, 40
Orthogneiss, 33, 97, 146

P
Palazzo Vertemate-Franchi, 1, 80, 81
Pangaea, 9, 69, 93, 147
Paragneiss, 33, 97, 107, 146
Parco Geologico Chiareggio, 119
Pass da Casnil, 104
Pass da la Duana, 97
Passo Cacciabella Sud, 104
Passo Tonale, 31
Peat bog, 35, 36, 71–73, 96
Pedroni-Del Prà, Bivacco, 68, 117
Pegmatite, 31, 32, 34, 76, 90, 103, 109, 120, 146
Pelitic schist, 71, 91, 97, 99, 107, 121, 147
Pennine nappes, 11–13, 19, 30, 33, 75, 89, 93, 95, 96, 107, 119, 145, 147
Pentlandite, 57
Phase diagram plagioclase, 44
Piazza Rodolfo Pestalozzi, 83, 84
Piedmontite, 102
Pietra ollare, 52, 80, 83, 146
Pillow structures, 30, 89, 90
Pinguicula, 35, 71, 96
Piuro (Plurs) landslide 1618, 3, 48, 49, 80
Plagioclase, 27, 29, 31, 33, 37, 40, 41, 43, 44, 71
Plan Canin, 45, 56, 86
Plän Vest, 98
Plate tectonics, 9
Platta nappe, 11, 73, 89, 91, 93, 113
Plaun da Lej, 55, 93
Pleistocene, 9
Plutonic rocks, 30, 31
Polished surfaces, 47, 81
Polymorphism, 33, 41
Prasinite, 11, 73, 75, 148
Precipitation, 59
Prehistoric incisions, 81
Promontogno, 52, 56, 57, 100
Prosto, 80
PT phase diagram Al_2SiO_5, 42
Pyrite, 23, 24, 88, 113

Pyrrhotite, 57

Q
Quarries, 33, 48, 52–57, 100, 118–120
Quarries, Tambo gneiss, 33, 52–57
Quartz, 24, 28
Quartzite, 3, 9, 34, 79, 91, 97, 102, 115
Quaternary, 9

R
Radiolaria, 30, 96, 148
Rhodochrosite, 102
Rhodonite, 26, 91, 102
Riebeckite, 26, 73, 95
Riss/Würm, 45
Rock hammer, 67
Rocks, 30–37, 78–79
Roticcio, 78, 100, 113

S
Salacina, Piz, 65, 107
San Fedelino granite, 31, 118, 147
San Giacomo Filippo, 83
San Lorenzo, Chiavenna, 83, 84
Santuario di Gallivaggio, 85
Sapphirine, 15, 25, 116, 118
Sasc da Corn, 93
Sasc Furä, Capanna, SAC, 68, 112
Sasc Taca, Stampa, 47–48
Sasso Dragone, 82
Saussurite, 95
Scapolite, 105
Scheuchzer, Johann Jakob, 50, 54, 55
Schists, 33, 76
Sciora, Capanna, SAC, 68, 110
Sciora group, 110
Sedimentary rocks, 30, 148
Sediments, 35–37
Segantini, Giovanni, 1, 72, 148
Sella del Forno, 89
Septimer Pass, 73
Serizzo (tonalite), 31, 104
Serpentine, 33

Serpentinite, 11, 29, 33, 34, 52, 73, 78, 91–96, 101, 119–120, 149
Sillimanite, 4, 25, 41–43, 109, 116, 121
Sissone, Val, 119–121
Snow, 59
Soapstone (Lavez), 52, 80, 146
Soglio, 1, 33, 56, 57, 112, 115
Sondrio, 119, 121
Staub, Rudolf, 4, 14, 149
Staurolite, 25, 97
Stilpnomelane, 26, 41, 43, 73, 94, 95
Studer, Bernhard, 4, 95, 116, 149
Suretta augengneiss, 97, 114
Suretta nappe, 11, 12, 13, 14, 93, 97, 101–102, 113, 115
SwissTopo App, 67
Swiss topographic maps, 65

T
Talc, 26, 52, 109, 112, 121
Talus debris, 30, 35, 44, 104
Tambo gneiss, 38, 56, 57, 82, 83, 85
Tambo granite, 56, 83
Tambo nappe, 11, 12, 13, 14, 83, 98, 107, 115
Tartaglione Crispo, Rifugio, 68, 118
Tectonic movements, 11–15
Tectonic structure of Alps, 12
Tectonic units, 12, 13, 144
Tegiola, Bocchetta, 116
Tethys ocean, 9, 33, 91, 93, 119, 149
Theobald, Gottfried, 4
Thin sections
    amphibolite with Ab-An, 41
    Al_2SiO_5-schist, 41
    Bergell granite, 41
    Chloritoid schist, 99
    Radiolaria, 94
    Riebeckite schist, 94
    Stilpnomelane schist, 41
    Ultramylonite, 117
Thunderstorms, 67
Titanite, 25, 28
Tombal, 40, 98–99, 100
Tombal kiln, 55, 98

Tonalite, 16, 31, 78, 104, 110, 121, 149
Tonalite-amphibolite contact, 121
Torre Belvedere, 72, 74
Tourmaline, 25, 28
Travertine, 75, 149
Tremolite, 26, 29, 58, 88, 105, 121
Triangia tonalite, 31, 119
Triassic, 9, 93, 115
Triassic marble, 75, 91, 97–99, 101
Triassic quartzite, 34, 61, 97, 99
Triple Point, 41, 42, 92, 107, 149
    Africa/Europe/Bergell granite, 3, 107
    Al_2SiO_5, 4, 41–43, 109
    Rhine/Danube/Po, 3, 44, 91
Trubinasca, 110

U
Ultramafic breccia, 110, 111, 116
Ultramafic Chiavenna zone, 52, 82, 109, 116
Ultramafic rocks, 48, 108–110, 146, 149
Ultramylonite, 116, 117
Uschione, 82

V
Val Albigna reservoir, 59
Val Codera, 4, 116
Val Forno, 86–88
Valmalenco, 119–121
Val Sissone, 121
Valun da la Trubinasca, 110
Vesuvianite, 88
Via Panoramica, 112–115
Volcanic dikes, 31, 32, 73, 91, 96, 100–101

W
Water, 3, 37, 53, 58–60, 89, 91, 95, 113
Waypoints, 67
Wegener, Alfred, 9
Wollastonite, 26, 105, 121
Würm, 44, 146

X
Xenoliths, 72, 88, 104, 112, 149
Xenolith swarms, 86, 87, 110

Y
Young granites, 3, 31, 37

Z
Zircon, 7, 25, 28, 118
Zircon U/Pb ages, 7, 118

SPRINGER NATURE

GPSR Compliance

The European Union's (EU) General Product Safety Regulation (GPSR) is a set of rules that requires consumer products to be safe and our obligations to ensure this.

If you have any concerns about our products, you can contact us on ProductSafety@springernature.com

In case Publisher is established outside the EU, the EU authorized representative is:

Springer Nature Customer Service Center GmbH
Europaplatz 3
69115 Heidelberg, Germany

The manufacturer's authorised representative in the EU is Springer Nature Customer Service Centre GmbH, Europaplatz 3, 69115 Heidelberg, Germany. If you have any concerns regarding our products, please contact ProductSafety@springernature.com

Printed and bound by CPI Group (UK) Ltd, Croydon, CR0 4YY
25/03/2026
02078177-0001